Lecture Notes in Computer Science   5183

Commenced Publication in 1973
Founding and Former Series Editors:
Gerhard Goos, Juris Hartmanis, and Jan van Leeuwen

Editorial Board

T0217341

Giuseppe Psaila   Roland Wagner (Eds.)

E-Commerce and Web Technologies

9th International Conference, EC-Web 2008
Turin, Italy, September 3-4, 2008
Proceedings

 Springer

Volume Editors

Giuseppe Psaila
Università degli Studi di Bergamo
Facoltà di Ingegneria
Viale Marconi 5, 24044 Dalmine (BG), Italy
E-mail: psaila@unibg.it

Roland Wagner
Johannes Kepler University Linz
Institute for Application Oriented Knowledge Processing (FAW)
Altenberger Straße 69, 4040 Linz, Austria
E-mail: rrwagner@faw.uni-linz.ac.at

Library of Congress Control Number: 2008933383

CR Subject Classification (1998): H.4, K.4.4, J.1, K.5, H.3, H.2, H.2.5, K.6.5

LNCS Sublibrary: SL 3 – Information Systems and Application, incl. Internet/Web and HCI

ISSN   0302-9743
ISBN 978-3-540-85716-7 Springer Berlin Heidelberg New York

Springer is a part of Springer Science+Business Media

springer.com

© Springer-Verlag Berlin Heidelberg 2008

Typesetting: Camera-ready by author, data conversion by Scientific Publishing Services, Chennai, India
Printed on acid-free paper   SPIN: 12466603   06/3180   5 4 3 2 1 0

Preface

The International Conference on E-commerce and Web Technologies (EC-Web) is a mature and well-established forum for researchers working in the area of electronic commerce and web technologies.

These are the proceedings of the ninth conference in the series, which, like previous EC-Web conferences, was co-located with DEXA, the International Conference on Database and Expert Systems Applications, which, this year, took place in Turin, Italy.

One key feature of EC-Web is its two-fold nature: it brings together both papers proposing technological solutions for e-commerce and the World Wide Web, and papers concerning the management of e-commerce, such as web marketing, the impact of e-commerce on business processes and organizations, the analysis of case studies, as well as social aspects of e-commerce (to understand the impact of e-commerce solutions on day-to-day life and the new opportunities that these behaviors open).

The technical program included 12 reviewed papers and two invited papers. Each paper was reviewed by five reviewers, in order to select only the best quality papers. The program included five sessions: "Security in E-Commerce" (with two papers), "Social Aspects of E-Commerce" (with two papers), "Business Process and EC Infrastructures" (with three papers), "Recommender Systems and E-Negotiations" (with four papers) and "Web Marketing and User Profiling" (with three papers).

We found the program interesting and we hope participants and readers feel the same. Furthermore, we hope the attendees enjoyed the conference and Turin.

June 2008

Giuseppe Psaila
Roland R. Wagner

Organization

Program Committee Chairpersons

Giuseppe Psaila   University of Bergamo, Italy
Roland Wagner   FAW, University of Linz, Austria

Program Committee

Marco Aiello   Rijksuniversiteit Groningen, The Netherlands
Esma Aïmeur   University of Montreal, CA
Damminda Alahakoon   Monash University, Australia
Sergio Alonso   University of Granada, Spain
Jörn Altmann   Seoul National University, South Korea and Intl.
                    University of Bruchsal, Germany
Manish Bhide   IBM India Research Lab, India
Sami Bhiri   DERI, Ireland
Sourav S. Bhowmick   Nanyang Technological University, Singapore
Susanne Boll   University of Oldenburg, Germany
Stephane Bressan   National University of Singapore, Singapore
Julián Briz   Universidad Politécnica de Madrid, Spain
Michelangelo Ceci   University of Bari, Italy
Wojciech Cellary   Poznan University of Economics, Poland
Francisco Chiclana   De Montfort University, UK
Byron Choi   Nanyang Technological University, Singapore
Jen-Yao Chung   IBM Watson Research Center, USA
Emmanuel Coquery   University Lyon 1, France
Arthur I. Csetenyi   Budapest Corvinus University, Hungary
Alfredo Cuzzocrea   University of Calabria, Italy
Radoslav Delina   Technical University of Kosice, Slovakia
Alexander Delteil   Orange Labs - France Télécom, France
Tommaso Di Noia   Politecnico di Bari, Italy
Petr Doucek   Prague University of Economics, Czech Republic
Schahram Dustdar   Vienna University of Technology, Austria
Johann Eder   University of Klagenfurt, Austria
Maria Jose Escalona   Universidad de Sevilla, Spain
Torsten Eymann   University of Bayreuth, Germany
Eduardo Fernandez   Florida Atlantic University, USA
Gianluigi Ferrari   University of Pisa, Italy
Elena Ferrari   University of Insubria, Italy
George Feuerlicht   University of Technology Sydney, Australia
Ludger Fiege   Siemens, Germany

Carlos Flavian	University of Zaragoza, Spain
Farshad Fotouhi	Wayne State University, USA
Eduard Cristóbal Fransi	University of Lleida, Spain
Yongjian Fu	Cleveland State University, USA
Walid Gaaloul	DERI, Ireland
Stephane Gagnon	Université du Québec en Outaouais (UQO), Canada
Jing Gao	University of South Australia, Australia
Piotr Gawrysiak	Supermedia, Poland
Peter Geczy	AIST - National Institute of Advanced Industrial Science and Technology, Japan
Chanan Glezer	Ben Gurion University, Israel
Claude Godart	University of Nancy & INRIA, France
Adnene Guabtni	University of New South Wales, Australia
Mohand-Said Hacid	University Lyon 1, France
G. Harindranath	Royal Holloway, University of London, UK
Aboul Ella Hassanien	Kuwait University, Kuwait
Josef Herget	University of Chur, Switzerland
Enrique Herrera-Viedma	University of Granada, Spain
Yigal Hoffner	IBM Zurich Research Lab., Switzerland
Birgit Hofreiter	University of Technology, Sidney, Australia
Christian Huemer	Vienna University of Technology, Austria
Michael C. Jaeger	Berlin University of Technology, Germany
Fabian Kaiser	Stuttgart University, Germany
Dimka Karastoyanova	University of Stuttgart, Germany
Gregory E. Kersten	Concordia University Montreal, Canada
Hiroyuki Kitagawa	University of Tsukuba, Japan
Jan Klas	University of Economics, Prague, Czech Republic
Gabriele Kotsis	Johannes Kepler University Linz, Austria
Sandeep Krishnamurthy	University of Washington, USA
Anton Lavrin	Technical University of Kosice, Slovakia
Juhnyoung Lee	IBM Watson Research Center, USA
Joerg Leukel	University of Hohenheim, Germany
Philipp Liegl	Technical University of Vienna, Austria
Leszek T. Lilien	Western Michigan University, USA
Antonio Gabriel Lopez	University of Jaen, Spain
Heiko Ludwig	IBM Watson Research Center, USA
Olivera Marjanovic	University of Sydney, Australia
Mário Marques Freire	University of Beira Interior, Portugal
Luis Martínez Lopez	University of Jaen, Spain
Massimo Mecella	University of Rome La Sapienza, Italy
Ralph Mietzner	University of Stuttgart, Germany
Bamshad Mobasher	DePaul University, USA
Carlos Molina-Jimenez	University of Newcastle upon Tyne, UK
Gero Muehl	TU Berlin, Germany
Guenter Mueller	University of Freiburg, Germany
Mieczyslaw Muraszkiewicz	Warsaw University of Technology, Poland

Dirk Neumann	University of Karlsruhe, Germany
Wee-Keong Ng	Nanyang Tech. University, Singapore
Ota Novotny	Prague University of Economics, Czech Republic
Anne-Marie Oostveen	Oxford Internet Institute, UK
Rolf Oppliger	eSECURITY Technologies, Switzerland
Stefano Paraboschi	University of Bergamo, Italy
Jan Paralic	Technical University of Kosice, Slovakia
Oscar Pastor	Valencia University of Technology, Spain
Vicente Pelechano	Techical University of Valencia, Spain
Günther Pernul	University of Regensburg, Germany
Ilia Petrov	Technische Universität Darmstadt, Germany
Ivana Podnar Zarko	FER, University of Zagreb, Croatia
Birgit Proell	Johannes Kepler University Linz, Austria
Gerald Quirchmayr	University of Vienna, Austria
Azzurra Ragone	University of Michigan, USA
Indrakshi Ray	Colorado State University, USA
Werner Retschitzegger	Johannes Kepler University Linz, Austria
Inmaculada Rodríguez-Ardura	Universitat Oberta de Catalunya, Spain
Michele Ruta	Politecnico di Bari, Italy
Henryk Rybinski	Warsaw University of Technology, Poland
Jarogniew Rykowski	Poznan University of Economics, Poland
Tomas Sabol	Technical University of Kosice, Slovakia
Paolo Salvaneschi	University of Bergamo, Italy
Farzad Sanati	University of Technology, Sydney, Australia
Nandlal L. Sarda	Indian Institute of Tech. Bombay, India
Yucel Saygin	Sabanci University, Turkey
Thorsten Scheibler	University of Stuttgart, Germany
Eusebio Scornavacca	Victoria University of Wellington, New Zealand
Steffen Staab	University of Koblenz, Germany
Jens Strueker	University of Freiburg, Germany
Aixin Sun	Nanyang Technological University, Singapore
Junichi Suzuki	University of Massachusetts, Boston, USA
Roger M. Tagg	University of South Australia, Australia
Samir Tata	GET/INT CNRS Samovar, France
Stephanie Teufel	University of Fribourg, Switzerland
Tobias Unger	Stuttgart University, Germany
Bartel Van de Walle	Tilburg University, The Netherlands
Willem Jan van den Heuvel	Tilburg University, The Netherlands
Krishnamurthy Vidyasankar	Memorial University, St. John's, Canada
Emilija Vuksanovic	University of Kragujevac, Serbia
Hannes Werthner	Technical University of Vienna, Austria
Branimir Wetzstein	University of Stuttgart, Germany
Janusz Wielki	Opole University of Technology, Poland
Marco Zapletal	Technical University of Vienna, Austria

Acknowledgement

The work was supported by the PRIN 2006 program of the Italian Ministry of Research, within the project "Basi di dati crittografate" (2006099978)

External Reviewers

Gabriela Vulcu
Nathalie Aquino
Ignacio Panach
Beatriz Marín
Giovanni Giachetti
Sergio España
Francisco Valverde
Nelly C. Fernández
Ludwig Fuchs
Christian Broser
Manuel Resinas

Table of Contents

Session 5 – Web Marketing and User Profiling

Secure Communication between Web Browsers and NFC Targets by the Example of an e-Ticketing System

Gerald Madlmayr[1], Peter Kleebauer[1], Josef Langer[1], and Josef Scharinger[2]

[1] University of Applied Sciences Hagenberg
{gmadlmay,cms02006,jlanger}@fh-hagenberg.at
[2] Johannes Kepler University Linz
josef.scharinger@jku.at

Abstract. Near Field Communication (NFC) is a radio frequency (RF) based proximity coupling technology allowing transactions within a range of up to 10 cm. Using NFC technology for transactions like payment or ticketing in the real world brings a great benefit in terms of time savings, usability and process optimization. Therefore we propose an e-ticketing system making use of this proximity technology especially focusing on security aspects of the system as well as the distribution of the tickets.

While other systems rely on ticket distribution via SMS or home-printing a paper ticket, our approach is based on a browser plug-in in combination with a contactless RFID reader at the client side. This installation is used to transfer the e-ticket from a ticket server to the user's PC client and to write the ticket over the proximity interface into the secure element of the NFC target. Thus an NFC target, a contactless smartcard or an NFC enabled mobile phone, can be used as a secure token. With this implementation we are able to bridge the gap between electronic internet transactions and the physical world in a secure way. Also the validation of the ticket at the point-of-access is based on this contactless technology. Our findings provide practical implications to implement web applications using NFC technology successfully.

1 Introduction

RFID (Radio Frequency Identification) is used by many daily used applications. For the consumer unnoticed and simple to use, they offer a popular alternative to conventional communication channels. Starting with simple access control possibilities up to complex data memories, different applications can be realized. A further development is represented by Near Field Communication (NFC) [1], a technology for a fast and simple exchange of small amounts of data. It opens new perspectives regarding the application development on mobile phones and other devices. Meanwhile NFC has found introduction to mobile phones, one of the most common means of communication [2].

G. Psaila and R. Wagner (Eds.): EC-Web 2008, LNCS 5183, pp. 1–10, 2008.

ABI Research [3] forecasts that by 2010 350 million mobile phones can be used as an RFID smartcard, an RFID reader/writer or a device which is able to establish an RF based peer to peer connection with other NFC devices [4]. While logistics and health care make use of long-range RFID technology, NFC relies on the smartcard standard ISO14443 [5] allowing wireless transactions only over a distance of up to 10 centimeters. This is part of the *Touch and Go* philosophy giving the user a new dimension of usability.

However, the use of proximity technology is popular at the point-of-sale or access gates. Now our approach is, to use this technology to a establish a secure communication between such an NFC target, a contactless smartcard or an NFC enabled mobile phone, and a web application. Various use cases for such a system, like user authentication at a website through a SIM card or payment in an online shop with a contactless creditcard, are considered. In the following we line out the implementation of our approach by the example of an e-ticketing system.

Because of transmissions on unsecured communication ways, suitable safety precautions for the data transfer must be met. Although proximity connections are difficult to compromise due to the short communication distance, different security measurements must be considered [6]. This refers to both, the integrity and authenticity of the transferred data. Suitable authorizing mechanisms are needed, so that the information sent cannot be queried by unauthorized parties. These requirements are not fulfilled for example by the common used communication protocols as shown in [7]. Therefore the new developed application must take care of providing secure data communication.

The related work section provides background information on ticketing systems, followed by a system overview of our implementation in section 3. Section 4 lines out the different components and parties involved and describes the communication flow. The major part of our implementation, a secure protocol for ticketing distribution, is given in section 5. The conclusion in the ending section lines out the pros and cons and deals with privacy concerns as well.

2   Related Work

Due to the implementation of NFC technology into mobile devices, e-ticketing systems are extensively discussed with regard to changes in processes. At the moment one distinguishes between *classic e-ticketing* and *AFC (automated fare collection)* [8]. In AFC systems the user checks-in before and checks-out after using the transport services. The system in the background automatically calculates the appropriate fare and credits it to the user's account. In case the transport services are used more often , the fare is adopted accordingly [9]. While AFC is interesting for public transport operators in mass transit, the classic e-ticketing will broaden its dominance in more individually cases like event or air travel ticketing. In terms of buying a ticket, surveys show that users are likely to spend small amounts of money using m-payment. Purchases resulting in a higher price are not yet popular in m-commerce. Although the booking of event or airline tickets through the internet is already well accepted by users [10], [11], [12].

At a gate, technologies allowing a user to quickly pass the gate result in a high consumer satisfaction. So far contactless technologies show the best consumer acceptance in this regard [13] and actually NFC enabled devices provide such an implementation. Therefore NFC is likely to be extensively used for ticketing and access in the near future. Other use cases for NFC are contactless payment and credit cards [14] as well as loyalty applications [15]. Whereas using different data bearers [16], like SMS or IP, for uploading data into the secure element (SE) [17] – also refered to as mobile trusted module (MTM) – are proposed, the possibility to use the contactless interface has not yet been considered.

Aigner and Dominikus propose the use of NFC devices for mobile couponing in [15]. Ordinary coupons are used to promote products and are distributed over the web or are paper-based ones. In these papers a new way of distribution through NFC is proposed. The coupons can be downloaded from a passive tag which is integrated into a poster or a newspaper. By touching the tag with an NFC enabled device, the coupon is downloaded into the SE of the mobile phone. With this mobile phone the user can then cash in the coupon at the cashier. The major focus of their work is based on securing the distributed infrastructure (tags) by the use of a PKI in order to prevent fraud and abuse. The implemented system uses a Windows Mobile PDA with a SD-NFC-Card.

Feng et al. present an SMS based ticketing system in [18]. Although the system uses Bluetooth communication to validate the ticket at the gate, the same system is suitable for an NFC solution as well. In this case also a PKI is used for authentication purpose between the different parties. The major benefit from NFC for this system would be the simple interaction at the gate as well as the use of the SE on order to store e-tickets or e-cash in a secure way. In the following we describe a system that allows the use of contactless smartcards or NFC targets for the interaction with a webapplication.

3 System Overview

The proposed system rests upon the fact that more expensive tickets are more commonly bought through means of e-commerce like the internet as through m-commerce. Additionally NFC enabled devices are becoming more and more popular for access control. Whereas today tickets bought through the internet might be home printed, delivered via mail or SMS, our approach is different. Having a plug-in in the browser that is able to communicate with a proximity reader at the client computer allows us to store and/or alter data in an NFC target within the range of this RFID reader (Fig. 1).

In order to do so, the user opens a website first, buys the ticket, and selects a delivery via NFC. Then – invisible to the user – a connection between the browser plug-in and the ticketing server is established. Additionally the plug-in opens a communication channel to the proximity reader and sets up an RF based

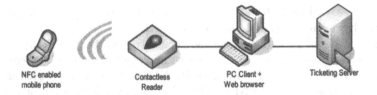

Fig. 1. The ticket server is able to communicate directly with the mobile phone through the browser and a contactless reader

communication with the NFC target followed by a ticket transfer from the server to the NFC target with end to end encryption.

Tickets are stored directly in the SE of the NFC target. For security reasons the emulation of the card usually is turned of and PIN-protected. This approach is already implemented in the phone used for the prototype (Nokia 6131 NFC) as well is proposed by the GSMA [19]. Additionally the system can be used for other purposes than ticketing as well. As a result of direct communication between web services and NFC devices, new use cases can be created:

- *Visualization and configuration of mobile phones:* Contacts or other configurations can be visualized and modified just by the use of the local web browser.
- *Authentication on websites:* The SIM (Subscriber Identity of Modules) can be consulted for the legitimacy of the access to websites. A further possibility would be the use of encrypting algorithms for authentication using secret keys which are stored on the cell phone [20]. NFC in this case can also be combined with other authentication mechanisms like PIN-Code or user Creditials for high security services (Mulit-Factor-Authentication).
- *Dynamic configuration of admission:* By the transmission of e.g. a secret, admission controls can be configured [21].
- *Payment:* Mobile payment systems can be combined with internet payment systems [13]. A web server communicates directly with an application in the secure memory, which makes the deduction of the amount.

The installation of the plug-in needs to be performed only once and then can be used by any arbitrary application. Without having the keys to authenticate against the SE of the NFC device, no 3rd party can get information out or alter information in the SE. The SE itself usually is configured by a mobile network operator (MNO) or a platform manager (PM) [19] in order to use it for contactless transactions. This process of configuration can be performed before issuing the devices but also through the mechanism of our system (post issuance). With regard to usability it can be said that the transmission of the data (e.g. tickets) itself is simple, secure and fast. These are key requirements for applications to be accepted by users.

4   Components

4.1   Public Key Infrastructure

For a safe communication a PKI (Public Key Infrastructure) presented in figure 2(a) is installed [22]. The control instance (MNO or PM) acts as a trusted authority and is equipped with a self signed certificate. It confirms the trustworthiness of subordinated instances by its signature. For this system the public keys need to be available to all instances on demand.

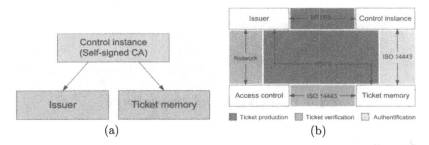

(a)   (b)

Fig. 2. Communication flow in the system (a) and PKI for an e-ticketing system (b)

4.2   Parties

For the realization of the system four different and independent parties are necessary. Different communication relations are established for the processes of purchasing a ticket, validating a ticket, and authenticating a ticket, which are shown in figure 2(b). This also depends on the following safety requirements, which are applicable for the connection types.

- *Access control:* The access control examines the authenticity of the tickets by communicating with the ticket memory in the SE. This instance is implemented as a contactless reader at the point-of-access.
- *Control instance:* The control instance confirms the validity of the ticket by its signature. This instance is a Certificate Authority and acts as a Trusted Third Party (TTP) in the architecture.
- *Issuer:* The issuer is responsible for the ticket generation and for the accounting. The instance possesses a pair of keys signed by the control instance. With these keys the issuer can authenticate himself against the control instance. The issuer actually is the instance selling the tickets to the consumer as well as validating the tickets at the access gate.
- *Ticket memory:* The ticket memory – also refered to as *traveler* – is an application for storing the tickets in a secure place as well as performing cryptographic functions for the ticketing system. We are using the JavaCard OS to implement the ticket memory in the SE of the NFC target. The ticket memory receives the ticket through the contactless reader at the client PC. The whole communication is secured by HTTPS. For authentication the ticket will be transmitted to the access control by using ISO14443.

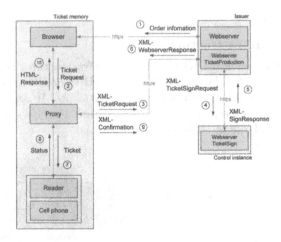

Fig. 3. Processes of the webticketing system

5   Protocol

5.1   Authentication of a Person

As shown in figure 4(b), for personal authentication the public key of the NFC device (PubKeyM) is transferred for encrypting the tickets. With the public key of the control instance the validity of the tickets can be verified by the ticket memory.

5.2   Ticket Production

Figure 3 shows an overview of the procedure of purchasing a ticket including the sequential steps of the protocol. This encloses the selection, the production, the signing, and the transmission of the ticket to the NFC device. XML based data structures are used for the communication among the components.

The data received by the browser is conveyed over a secured HTTPS connection to the web server (Step 1). Secondly the web browser is redirected to the plug-in of the browser (Step 2), which actually acts as a proxy. This proxy sets up an additional connection to the ticket production server requesting a ticket (Step 3). For the grant of the integrity and authenticity the issuer requests the control instance to sign the produced ticket (Step 4 and 5). The production of the ticket involves the generation of a pair of asymmetric keys whereas the public one is included in the ticket. Then the authorized issuer encrypts the ticket with the public key of the ticket memory. The validity is confirmed by the following signature. Starting from this time it can be only read by the application in the SE, since this is the only instance, which possesses the private key for decoding. After being transmitted to the proxy application (Step 6), the ticket is transfered by the reader over a proximity connection to the handset (Step 7). First the signature of the control instance is examined by the NFC device. With a valid signature and an error free operational sequence of the procedure the ticket can be stored in the SE. Then a confirmation message is sent to the issuer (Step 8 and 9) and a success message is displayed in the browser of the user (Step 10). In case of an error, the information is not stored in the NFC device and an appropriate message is sent back to the issuer and the web browser.

5.3   Ticket Verification

Another improvement of the whole ticketing process occurs during the transmission of the data to the reader device of the access instance. At this point it is verified, manipulated, and returned to the mobile equipment. This also presupposes the following precautions, that only authorized devices can request and select tickets from NFC devices:

- Authentication of the reader
- Encrypted communication
- Strictly scheduled protocol sequence
- Collection of protocol errors and protocol aborts

Based on these terms the communication protocol for the data exchange between the reader and the NFC device – shown in figure 4(a) – has been developed.

Protocol Sequence. After the selection of the NFC equipment the reader starts a protocol sequence by sending the identification number of the organizer (ID_Issuer) and a random number (R1). This is the only step in the protocol sequence, which is done encrypted transfer. Every other message is secured by asymmetrical encryption. The transmitted random numbers in each message are used for a fixed protocol sequence. Thus the protocol messages cannot be read by unauthorized persons or used later again (Replay attacks) [7]. With the issuer identification (Step 1) the NFC device can check whether there are existing tickets for this issuer or not. All applicable TicketIDs are transmitted to the reader encrypted, using the public key of the issuer (IssuerKey) stored in the ticket. In step 3 the reader selects the correct ticket for the event by selection of the event identification number (ID_Event).

Due to the fact that for each ticket a unique asymmetrical key (TicketKey) is generated, the identification number of the ticket (ID_Ticket) is transferred first, encrypted with the public key of the issuer. With this number the reader can query the correct private key from a remote data base (online) or a white list directly stored in the reader (offline) for decoding the ticket.

The second part of the message contains all relevant ticket data, encrypted with the public key of the ticket. Now the data is verified and, if the ticket is still valid, the ticket is either sent to the NFC device modified (if the ticket can be used more than once) or is deleted (if the ticket can only be used once).

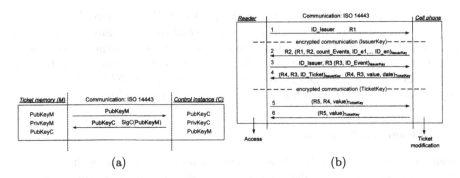

(a)   (b)

Fig. 4. Authentication of a person (a) and protocol sequence for ticket verification (b)

6   Implementation and Discussion

The ticket memory is implemented as a JavaCard Applet in the SE of a Nokia 3220 (NFC-Edition) or a Nokia 6131 NFC with an additional J2ME application to view the content of the ticket memory. On the client side (PC) there is a proxy running, to establish a communication between the contactless reader and the server of the issuer. We used the PC/SC interface of the contactless reader (Omnikey 5321) for communication with the NFC target. The proxy was implemented in Java using the *javax.smartcardio* package for the PC/SC communication. The proxy itself does not implemented means of security as the connection between the NFC target and the issuer is end-to-end encrypted. Thus man-in-the-middle attacks to retrieve data are not feasible. The issuer is implemented using a Tomcat web server. The NFC target stores the credentials for the authentication in the SE whereas the issuer holds this information in the JRE Certificate store.

Although the communication and data is pretty well protected, there is still a target for attackers. The installation of such a proxy on a client PC is still a security risk, as it could contain malicious code. Thus only the installation of signed applications should be considered. The PC system on the client side is the achilles' heel of the system. Thus the data is end-to-end encrypted between the smartcard and the issuer's system. Without a bilateral authentication, neither the smartcard nor the server application will allow a transaction. However, an attacker might block the system from working and thus cause inconvenience for both the consumer and the service provider.

7   Conclusion

The e-ticketing system presented in this paper shows that the NFC technology acts as an enabler for new means of distributing e-tickets. The technology provides a good usability to the end consumer beside ensuring high security for the data transmitted and stored in the SE. NFC is a good choice for ticketing in terms of delivery and validation at the gate. A separate hardware (contactless reader at the client PC) as well as a plug-in for the browser is needed to use these services. The major benefit over using SMS as a data bearer is, that the issuer immediately knows whether the e-ticket arrived safely or not, which is not the case for SMS. For the consumers point of view, there is no additional task to take in order to use the handset as a secure token. The handset is also able to display the content of the SE. Due to the fact that there are already 20 million contactless creditcards issued in the United States [14] the popularity of such services will raise. The proposed setup can also be used to implement a payment system for such credit cards on a home PC. However, due to the fact that a further device/application has to be installed at the end consumer and because of similar communication technologies such as Blue Tooth or WiFi are already on the market, the acceptance cannot be guaranteed.

7.1 Privacy

During the design of the system special attention was payed to the privacy issues. At no point a 3rd party can access the information in the SE without holding the correct key in order to access or decrypt the information stored there. Even authorized instances are not able to read other ticket information than their own. Additionally the serial number of the SE – used in contactless communication (T=CL) for anticollision purpose – is neither used for authentication or identification nor involved in the ticket production or verification process. From a technical point of view this is especially necessary as also a random serial number can be used by the mobile device during the anticollision process (as specified in NFCIP-1).

References

1. International Organization for Standardization: Near Field Communication - Interface and Protocol (NFCIP-1). ISO/IEC 18092 (2004)
2. Madlmayr, G., Ecker, J., Langer, J., Scharinger, J.: Near field communication: State of standardization. In: Michahelles, F. (ed.) Proceedings of the International Conference on the Internet of Things 2008, ETH Zürich, vol. 1(1), p. 6 (03 2008)
3. ABI Research: Near Field Communications (NFC) - Leveraging Contactless for Mobile Payments, Content and Access. Research Report (01 2007) Report Code: RR-NFC
4. Kunkat, H.: NFC und seine Pluspunkte. Electronic Wireless 01, 4–8 (2005)
5. International Organization for Standardization: Proximity cards. ISO/IEC 14443 (2003)
6. Hancke, G.P.: A Practical Relay Attack on ISO 14443 Proximity Cards. Technical report, University of Cambridge Computer Laboratory (2005)
   http://www.cl.cam.ac.uk/~gh275/relay.pdf
7. Heydt-Benjamin, T.S., Bailey, D.V., Fu, K., Juels, A., O'Hare, T.: Vulnerabilities in first-generation RFID-enabled credit cards. In: FC 2007, vol. 11, pp. 1–22 (2007)
8. Stroh, S., Schneiderbauer, D., Amling, S., Kreft, C.: Next Generation eTicketing, 1st edn. Booz Allen Hamilton (01 2007)

9. Transport for London: The oyster card (02 2008) (last visited, 02/27/2008), http://www.tfl.gov.uk/
10. Xu, H., Teo, H.H., Wang, H.: Foundations of SMS commerce success: lessons from SMS messaging and co-opetition. HICSS, 90 (01 2003)
11. Mallat, N., Rossi, M., Tuunainen, V.K., rni, A.: The impact of use situation and mobility on the acceptance of mobile ticketing services. HICSS 2, 42b (2006)
12. Mobile Electronic Transactions Ltd. Keilalahdentie 2-4, 02150 Finnland: MeT White Paper on Mobile Ticketing. 1.0 edn. (01 2003)
13. Zmijewska, A.: Evaluating Wireless Technologies in Mobile Payments - A Customer Centric Approach. In: Proceedings of the International Conference on Mobile Business (ICMB 2005), USA, vol. 04, pp. 354–362. IEEE Computer Society, Los Alamitos (2005)
14. Atkinson, J.: Contactless Credit Cards Consumer Report 2006 (04 2006), http://www.findcreditcards.org/
15. Aigner, M., Dominikus, S., Feldhofer, M.: A System of Secure Virtual Coupons Using NFC Technology. PerComW 5, 362–366 (2007)
16. Giesecke and Devrient Munich, Germany: White Paper: Bearer Independent Protocol (BIP). 1.0 edn. (2006)
17. Bishwajit, C., Juha, R.: Mobile Device Security Element. Mobey Forum, Satamaradankatu 3 B, 3rd floor 00020 Nordea, Helsinki/Finland (02 2005)
18. Feng, B., Anantharaman, L., Deng, R.: Design of portable mobile devices based e-payment system and e-ticketing system with digital signature. ICII 6, 7–12 (11 2001)
19. GSMA London Office 1st Floor, Mid City Place, 71 High Holborn, London WC1V 6EA, United Kingdom: mobile NFC technical guidelines. 2.0 edn. (04 2007) 1st Revision
20. SmartTrust Inc.: Whitepaper - Mobile Authentication. Revision: B edn. (02 2004) BD 04-0041
21. Su, S.L., Garg, H.: Designing SMS applications for public transport service system in Singapore. ICCS 2, 706–710 (2002)
22. Noll, J., Calvet, J.C.L., Myksvoll, K.: Admittance Services through Mobile Phone Short Messages. ICWMC 1, 77 (2006)

A Light Number-Generation Scheme for Feasible and Secure Credit-Card-Payment Solutions

Francesco Buccafurri and Gianluca Lax

DIMET, University of Reggio Calabria
via Graziella, Località Feo di Vito, 89122 Reggio Calabria, Italy
bucca@unirc.it, lax@unirc.it

Abstract. Disposable-number credit card is a recent approach to contrasting the severe problem of credit card fraud, nowadays constantly growing, especially in credit-card-based e-commerce payments. Whenever the solutions cannot rely on a secure extra communication channel between cardholder and issuer, the only possibility is to generate new numbers on the basis of some common scheme, starting from secret shared initial information. However, in order to make the approach feasible, the computational load both on issuer and customer side should be minimized, also to reduce the cost of user-side devices, keeping yet an adequate security level. In this paper we present a disposable-number credit card scheme meeting the above goals, going a step ahead w.r.t. the state of the art.

1 Introduction

Credit card fraud is nowadays a serious problem whose dimension is constantly growing. Indeed, there are a number of techniques used by attackers to sniff (during both on-line and traditional transactions) the fixed credit card number used for authentication, and to use it for fraudulent payments. This is of course a direct consequence of the intrinsic weakness of the traditional credit card processing system, where the key used for authentication is long-term, semi-secret, transmitted over insecure channels, sometimes completely disclosed. Consider that credit cards are still widely used in e-commerce (as well as in other e-activities) since it often happens that alternative secure payment methods (like [15,10]) are not applicable or preferred.

A recent approach to contrasting the above problem is based on the concept of disposable-number credit cards. According to this scheme, issuer and customer agree on a number to use for the transaction, then they discard it, generate a new number for the next transaction and so on. This way, sniffing an authentication number during a transaction does not give the attacker any useful credential.

There are commercial [9,8] solutions based on the above scheme. Unfortunately, such solutions are either still insecure, when the cardholder gets the disposable number from the issuer Web site and authenticates itself by sensible data (like a standard credit card number) [16] or too expensive (and little friendly), when the generation of the new number is executed on board of a smart

G. Psaila and R. Wagner (Eds.): EC-Web 2008, LNCS 5183, pp. 11–20, 2008.

card, and the issuer provides the cardholder with an additional device capable of displaying the disposable number. In principle, the extra cost related to the additional device could be eliminated by exploiting simple software solutions to display disposable numbers, interfaced with standard smart-card drivers. Anyway, the cost of (secure) smart cards is not irrelevant, so that a real applicability of the above strategy could be strongly related with the possibility of decreasing significantly also the cost of the card. Another related problem is that the number generation scheme should be enough secure, because, differently from schemes based on an extra communication channel (typically the Web), where numbers can be generated randomly by the issuer, necessarily there will be a mathematical link between a given disposable number and the successor one. This mathematical link could be thus exploited by the attacker in order to predict new numbers on the basis of the past ones. As a consequence adequate efforts are necessary in order to design number generation schemes sufficiently robust, w.r.t. possible attacks. There are a number of research solutions [17,5] addressing the above problem. The first approach of this type is presented in [13] that generates a new authentication number by encrypting in a smart card a set of possible restrictions describing some elements of the transaction itself. Starting from the consideration that encryption is too expensive to be realistically used in this context, the authors of [17] propose the use of context free grammars (thus not relying on any cryptographic algorithm) to generate disposable credit card numbers. Context free grammars present the property that the generation and validation of strings belonging to a given language can be done in polynomial time, but it is unfeasible to find the grammar given only the strings generated by it, since any conjectured grammar may fail on a new input string. However, the authors do not give any suggestion about how context free grammars have to be generated. As a consequence, an unlucky generation of the grammar may allow an attacker to easily guess the grammar. Moreover, as stated in [18], there exists no theoretical result about how difficult is it to guess another string which belongs to the same language, thus showing the impossibility to guarantee the security of this technique. Also the authors of [5] propose a more efficient solution using cryptographic hash functions rather than encryption.

In this paper we do a step ahead. Indeed, we propose an authentication number generation scheme that is much computationally easier than previous approaches. It is based on a new fast non-cryptographic hash function as the core of the solution, and is highly secure too. The interesting thing is that the computation which the generation scheme is based on is implementable in hardware by very simple circuits whose cost is relevantly smaller than the traditional smart-card chip. Observe that the high computational lightness is a key issue also from the perspective of the issuer-side computational load. This is not irrelevant whether we think of a huge number of clients simultaneously using their credit card.

The rest of the paper is organized as follows. In the next section we present the conceptual basis of our proposal. The detailed definition of the approach is given in Section 3, where we deal with the elements composing the disposable-number

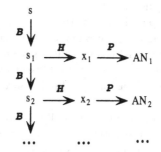

Fig. 1. Number generation scheme

generation scheme. Section 4 deals with the security issues. An implementation of our proposal is provided in Section 5. Finally, in Section 6 we draw our conclusions. For space limitations the theorem proofs are not included in the paper.

2   Overview of the Proposal

The number generation scheme is based on the following elements:

1. an initial seed s, generated by the credit card issuer, consisting of a k-bit string;
2. a *basic* function $\mathcal{B}$ to obtain a k-bit string from another k-bit string;
3. a hash function $\mathcal{H}$;
4. a *projection* function $\mathcal{P}$;

The scheme to generate disposable numbers is shown in Figure 1. In particular, by computing $\mathcal{B}(s)$ we obtain a new seed s_1 that is recorded in place of s. Starting from s, it is possible to create a chain of values $s_1, ..., s_n$ such that $s_{i+1} = \mathcal{B}(s_i)$ for $i > 1$. Since the function $\mathcal{B}$ is reversible, such a chain can be traversed also backward, by considering that $s_i = \mathcal{B}^{-1}(s_{i+1})$. Thanks to such a feature, the chain values are not required to be pre-computed and stored, but they can be generated on the fly (the advantages resulting from this feature regard all the cases when the issuer has to check the validity of an already burnt number, like in case of refunds).

Once the new seed s_1 is generated, we compute $x_1 = \mathcal{H}(s_1)$. The one-way property of the hash function guarantees that the knowledge of x_1 does not give an attacker the possibility to guess s_1.

The last step, $AN_1 = \mathcal{P}(x_1)$ is necessary to transform the value obtained by the hash function to a new credit card number. The following elements s_i, such that $i > 1$, and the corresponding ANs are obtained by iterating the above procedure.

An important issue regards the hash function to be used. We need to exploit its *one-way* property and this is the only property we required. On the contrary, the approach followed in [4] uses SHA-1 [7], a particular hash function provided with additional features (it must satisfy the *collision-resistant* property) and for this reason defined *cryptographic* hash function.

In our approach, this strong assumption is not necessary. We just require that given $x = \mathcal{H}(s)$, it is hard for the attacker to guess s from the knowledge of x (i.e., to invert the function by a brute-force attack). Observe that for a cryptographic hash function it is required the infeasibility of finding any y (even different from s) such that $x = \mathcal{H}(y)$.

In order to reach the goal of computationally *non-invertibility* (by generate and test approaches) of the hash function, a possible strategy is to design a hash function producing a large number of collisions and, thus, a sufficiently large research space. In our scheme, we can easily set parameters in order to obtain about 2^{450} collisions per value.

Note that our approach does not consist simply in the substitution of a weak hash function in place of a cryptographic one in a typical number generation scheme (like [5]). It is intuitive to understand that this would result in a very insecure approach just because of the weakness of the hash function itself (and because, in a scheme like [5], a new number is calculated as a hash of a seed that contains the number obtained at the previous step). We have designed thus a new weak hash function and, coherently, a new number generation scheme guaranteeing the security of the approach.

3   Number Generation Scheme

In this section we give the definitions of the elements composing the number generation scheme, that are the *basic function* $\mathcal{B}$, the *hash function* $\mathcal{H}$, and the *projection function* $\mathcal{P}$, we study some important properties of the functions $\mathcal{B}$ and $\mathcal{H}$, and, finally, we deal with the problem of the definition of the *initial seed* s. Observe that the choice of the basic function, the hash function and the projection function, cannot proceed orthogonally. Since the non-secret result is the composition of the three functions, we have to avoid that they are based on the same elementary operations, giving useful information to the attacker to proceed by crypto-analysis techniques. To prevent this, as we will explain in the following in this section, the basic function is based on string reverse and sum, the hash function is based on XOR and shift, and the projection function is based on scaling operations.

Before going into detail about the elements composing our scheme, we need some preliminary notations that will be used along the paper.

Notations. We denote by $x^k = (x_0, \ldots, x_{k-1})$ a k-bit string, where x_j, such that $0 \leq j \leq k - 1$, represents the j-th bit. We denote by $\widetilde{x}^k = (x_{k-1}, \ldots, x_0)$ the k-bit-reverse string. Moreover, $x^k + 1$ denotes the k-bit string representing the number obtained by summing x^k thought as a binary number and 1, in 2^k-modulo arithmetic. For example, given $x^3 = 111$, $x^3 + 1$ represents the string 000, since $(111 + 001)(\mathbf{mod}\ 1000) = 000(\mathbf{mod}\ 1000)$. Moreover, we denote by

1^k (0^k, resp.) the k-bit string composed of all 1s (0s, resp.). Finally, let x^j be a j-bit string. We denote by $x^i x^j$ the $(i+j)$-bit string obtained by juxtaposing x^j to x^i.

3.1   The Basic Function

The basic function $\mathcal{B}$ allows us to generate the sequence of seeds used in the scheme. The function is defined as follows.

Definition 1. *Given a k-bit string s^k, then $\mathcal{B}(s^k) = \widetilde{s}^k + 1$.*

In words, $\mathcal{B}(s^k)$ is obtained by reversing the string s^k and, then, by summing 1 (modulo 2^k). This function allows us to have a new seed at each generation. Clearly, the period of this function should be as large as possible, hopefully 2^k (the upper bound), in order to have a negligible probability of re-generating a seed during a plausible life of a credit card. The next theorem ensures that the above goal is reached provided that k is odd.

Theorem 1. *Given a k-bit string r_0^k with $k \bmod 2 \neq 0$, let R^k be the sequence $\langle r_0^k, \ldots, r_{2^k-1}^k \rangle$ such that $r_i^k = \mathcal{B}(r_{i-1}^k)$ for $1 \leq i \leq 2^k - 1$.*
Then, it holds that $r_i^k \neq r_j^k$, for any i, j such that $0 \leq i < j \leq 2^k - 1$.

3.2   The Hash Function

As observed in Section 2, our proposal is based on the usage of a *weak* hash function. In particular, we refer to a hash function able to guarantee a *weak one-wayness*, obtained by the generation of a large number of collisions.

The first question is understanding if some already existing weak hash function can be used for our purpose. A good candidate could be CRC (Cyclic Redundancy Check) [3], a non-cryptographic hash function that is widely used in error-detection contexts, both for its effectiveness to detect many kinds of errors and for its efficiency, since a simple shift register circuit can be constructed to compute it in hardware [11]. Observe that CRC is much faster than cryptographic hash functions, even if it is computed in software[1]. However, we will see in Section 4 that CRC cannot be used in our scheme due to an intrinsic weakness which it suffers from. So we have to design a new weak hash function keeping the nice computational features of CRC but eliminating its weakness. Let us describe first how CRC works.

CRC is computed to produce a n-bit string, named *checksum*, starting from an arbitrary length string, called *frame*, such that also a slight change of the frame produces a different checksum. The checksum is computed as the rest of

---

[1] In order to test the efficiency of our proposal, we have performed some experiments comparing the efficiency of CRC (64 bits) computation with SHA-1. The experimental results show that CRC is one magnitude order faster than SHA-1. Indeed, computing 10^9 CRC hashes required about 300 seconds, whereas SHA-1 took about 3800 seconds.

the binary division with no carry bit (it is identical to XOR) of the frame, by a predefined *generator polynomial*, a $(n+1)$-bit string representing the coefficients of a polynomial with degree n. CRC is thus parametric w.r.t. the generator polynomial and for this reason there are many kinds of CRCs. For example, the most frequently used are CRC32 and CRC64 that generate a checksum of length 32 and 64 bits, respectively. Obviously, the higher the checksum length, the better the effectiveness of CRC in error detecting. Beside dependence on the generator-polynomial length, CRC is parametric w.r.t. the value of its coefficients. Consequently, the goodness of CRC strictly depends also on the latter parameter. Among the several existing CRCs, in the following we will refer to CRC64 whose generator polynomial is defined by the ECMA standard [2], since we argue that a 64-bit fully tested CRC offers satisfactory robustness features.

CRC satisfies the one-way requirement introduced in Section 2. Indeed, given a k-bit frame s^k (with $k > 64$) and its w-bit (with $w = 64$) checksum c^w computed by CRC, there are 2^{k-w} collisions, that is there exist 2^{k-w} k-bit strings s_i^k such that $CRC(s_i^k) = c^w$. We may vary k in order to increase the number of collisions generated by CRC to any value (for example 2^{450}) to the goal of making practically infeasible a brute-force attack attempting to find the original frame s^k. Moreover, its implementation easiness and efficiency make CRC very appealing to be used in this context.

Beside these nice features, CRC is not immune from malicious attacks exploiting its linearity w.r.t. XOR (this weakness has been widely documented in the literature and already exploited in some application contexts, like Wep [1,19]). In particular, it holds that $CRC(a \, XOR \, b) = CRC(a) \, XOR \, CRC(b)$, that is the checksum of the XOR of two numbers is equal to the XOR of the checksums of the two numbers. In our case, this property of CRC could be in principle exploited by an attacker to obtain the hash of the i-th seed of an user (i.e. $x_i = CRC(s_i^k)$) starting from the knowledge of (1) the hash of the j-th seed of the user and (2) the XOR between s_i^k and s_j^k (this attack is analyzed in Section 4).

We need thus to construct a hash function not suffering from the above problem, and preserving the other nice features of CRC. The idea is to apply a cyclic right shift to each seed before calculating the CRC value. But, clearly, the number of such shifts cannot be equal for each seed, otherwise the prediction described above can be identically applied. The solution we adopt is that the number of cyclic right shifts applied on a given seed s_i^k is equal to the number of 1s occurring in the seed itself. We denote by $\overrightarrow{s}_i^k$ the resulting k-bit string.

Now we are ready to define our hash function $\mathcal{H}$.

Definition 2. *Given a k-bit string s_i^k, then $\mathcal{H}(s_i^k) = CRC64(\overrightarrow{s}_i^k)$.*

3.3   The Projection Function

The numbers found on credit cards share a common numbering scheme. For a standard 16-digit credit card, the number consists of a single-digit major industry identifier (MII) (4 for Visa, 5 for MasterCard, and so on), a five-digit issuer identifier number (IIN), an account number (AN), and a single digit checksum

(C) computed by the Luhn algorithm [6]. Thus, given a major industry and a given issuer, for every users only digits from 7 to 15 can change. The projection function $\mathcal{P}$ transforms the 64-bit string x_i^{64} generated by $\mathcal{H}$ into a 9-digit number AN_i. Observe that a trivial implementation of such a function as $\mathcal{P}(x^{64}) = x^{64}$ **mod** 10^9 is not a good solution since (1) it cannot be easily realized via hardware because 10^9 is not a power of 2 and (2) the distribution of values so obtained is not uniform (values ranging from 0 to $(2^{64} - 1)$ **mod** 10^9 are more probable than the remaining ones).

To overcame this problem we have implemented the following solution. First, we introduce two notations. Given a k-bit string K, we denote by $[k]_{i,j}$ with $1 \leq i \leq j \leq k$ the sub-string of K obtained by keeping the $j - i + 1$ bits starting from the i-th left-most bit. For example, given $k = 1000$, $[k]_{1,2} = 10$. Given a decimal number number N, we denote by $[N]_j$ its j-th left-most digit. For example, given $N = 56789$, $[N]_2 = 6$.

Our solution requires a modulo-10 (decimal) counter C, initialized to 0. Initially, C is increased by 1 if the first bit (i.e. the most significant) of x^{64} is 1. The remaining 63 bits of x^{64} are partitioned in 9 buckets of 7 bits. The i-th bucket is used to set $[AN]_i$, that is the i-th digit of AN. The i-th 7-bit bucket may assume a value v_i ranging from 0 to 127, and is partitioned again in 11 intervals. The first 10 intervals have size 12, whereas the last one 8. Let p_i be the index of the interval which v_i belongs to. If p_i is less than 10, then the value of the i-th digit of AN is set to p_i. Otherwise (i.e., p_i is either 10 or 11), $[AN]_i$ is set to the value stored in C and C is increased by p_i (modulo 10). It is easy to show that this procedure results in a uniform distribution of ANs.

4 Security Issues

In this section we analyze the robustness of the proposed disposable number generation scheme with respect to a number of possible strategies followed by an attacker to guess the next credit card number. In our analysis we consider firstly the case of a brute force attack trying to find the current seed s_i starting from the sniffing of the last-used credit card numbers. Then we describe possible cryptanalytic attacks exploiting the knowledge of (more than one) consecutive credit card numbers. The analysis is done assuming that $k = 511$.

Consider a brute-force attack conducted knowing some credit card numbers used by the user (thus he knows some AN_i of our scheme). Clearly, he has to guess the source hash value (i.e. x_i of our scheme) starting from AN_i. Since the projection function maps in an uniform way all the 2^{64} x_i in the set of 10^9 AN_i, the probability of success is $(2^{64}/10^9)^{-1}$, that is about 2^{-34}. At this step the attacker has found 2^{34} potential x_i. Let suppose the attacker can detect the correct x_i among the potential 2^{34} values. Then, he must find the original seed s_i such that $\mathcal{H}(s_i) = x_i$. By repeating the above reasoning, we obtain that the attacker will find $2^{511}/2^{64} = 2^{447}$ potential solutions. Observe that, if the value $\bar{s}_i$ chosen by the attacker (among the 2^{447} found) differs from the actual s_i (i.e. the current seed of the fraud victim), then the probability that $\overline{AN}_{i+1} = AN_{i+1}$, where $\overline{AN}_{i+1}$ is the next credit card number obtained by $\bar{s}_i$ and AN_{i+1} is that one obtained by s_i, is $\frac{1}{10^9}$, that coincides with the probability of guessing a valid credit card number with no background knowledge. Thus, these results should discourage the attacker from trying to break the scheme by brute force attacks.

Now consider the case the attacker knows a sequence C of c consecutive credit card numbers spent by the victim. By a brute force the attacker should test $10^{c*9/2}$ seeds to find a seed $\bar{s}$ producing such a sequence C. Observe that, since our generation scheme produces a mapping between a set of 2^{511} bit strings and a set of 10^9 (i.e., about 2^{30}) numbers, till c is less than $511/30 - 1 \approx 16$, the probability of guessing also the next credit card number of the victim is again 10^{-9}. For higher c, this probability becomes 1 but the number of seeds to test is really too large (more than 10^{76}).

Finally, consider an attack based on the weakness of the CRC computation. In Section 3, we have noted that every two steps, the "noise" introduced by the reverse operation is *quasi*-cancelled. To understand how this could be exploited for an attack, we observe that when a seed s_i^k has both the left-most and the right-most bit 0 (i.e., every four steps), the attacker knows that s_i^k XOR $s_{i+2}^k = 10^{k-2}1$ (recall that, according to our preliminary notations, $10^{k-2}1$ denotes a k-bit string of the form $1\cdots1$, with $k-2$ 0s). Thus, the CRC of s_{i+2}^k is easily predictable by exploiting the above property. This behavior can be generalized also for other bit configurations. It is easy to see that if s_i^k is of the form $00\cdots01$, then we expect that the XOR with the seed generated two steps ahead is of the form $10^{k-3}11$. Again, if s_i^k is of the form $10\cdots00$, then we expect that the XOR with s_{i+2}^k is of the form $110^{k-3}1$. Finally, if s_i^k is of the form $10\cdots01$, then we expect that the XOR with s_{i+2}^k is of the form $110^{k-4}11$. This is a symptom of the alternating destructive effect of the reverse operation and, further, of the general invariance of the internal part of the seed, when the basic function is applied. Observe that this negative effect is maximum whenever the seed is palindromic, because the effect of the reverse is null also on a single step.

The next theorem gives us the probabilistic support that a quasi-random generation of the initial seed prevents this drawback for the entire credit card life.

Theorem 2. *Let t and k be two positive integers such that $t < 2^{\frac{k-4}{2}}$. Let s^k be a k-bit seed of the form $10c^jd^{k-4-2j}e^j00$, where c^j and e^j are j-bit strings, d^{k-4-2j} is a $(k-4-2j)$-bit string containing at least one 0 and $j = \lceil log_2t \rceil + 1$. It holds that the sequence $S^t = \langle s_0^k, \ldots, s_t^k \rangle$ such that $s_0^k = B(s^k)$ and $s_r^k = B(s_{r-1}^k)$ for $1 \le r \le t$ does not contain any seed of the form $10f^{k-4}01$, where f is a $(k-4)$-bit string.*

The theorem states that (i) fixing both the first and the last two bits of the initial seed (to 10 and 00, respectively), and (ii) ensuring that the seed contains an internal centered range whose bounds are distant $\lceil log_2t \rceil + 1$ from the bottom (and the top) of the seed itself such that at least one 0 occurs in this interval, then it results that for at least t applications of the basic function (thus, at least for the next t credit card transactions), we do not generate bad seeds (i.e., seeds of the form $10\cdots01$). For example, in order to have the above property for the first $t = 50.000$ transactions, it suffices to set the initial seed to $10s_1^{17}s^{k-38}s_2^{17}00$, where s_1^{17}, s_1^{17} and s^{k-38} are randomly generated, with the only constraint that s^{k-38} contains at least one 0. It is easy to verify that the probability that a randomly generated string s^{k-38} does not satisfies the above requirement is $\frac{1}{2^{k-38}}$ (thus the blind random generation could be also accepted). For example in the case $k = 511$ this probability is $\frac{1}{2^{473}}$.

5   Implementation Issues

In this section we sketch the design of the hardware device implementing the number generation scheme so far described, in order to make evident that a strong positive point of our proposal is its feasibility and cheapness (especially w.r.t. other approaches based on smart card).

A concrete protocol implementing our scheme requires that the initial seed is generated by the credit card issuer, that is the provider of the device itself. In the following we assume the seed length is $k = 511$, that, as analyzed in the previous sections, guarantees a high security level. However there is no serious difficulty from the hardware point of view in further increasing this value in order to further hardening the system.

The device is equipped with three circuits implementing the basic function, the hash function and the projection function. The circuit implementing the basic function is composed of a 512-bit shift register R, storing the current seed and allowing the shift operation, and an adder used to implement the increment operation (for space reasons, we cannot describe the circuit).

Concerning the hash function implementation, it requires a simple shift register circuit and XOR gates (for details about CRC implementation and its faster table-driven implementation see the wide related literature [11,12,14]).

Finally, the projection function is implemented by means of a small combinatorial circuit having 7-bit input and 4-bit output (for example a ROM of $2^7 \cdot 4 = 512$ bits). This circuit works on each of the 9 buckets of 7 bits as described in Section 3.3, and returns the variable 9-digit number AN that, together with the fixed major industry identifier MII, the identifier number IIN, and the checksum C, produces the final disposal credit card number.

6   Conclusions

In this paper we have proposed a new number generation scheme used for CCT credit cards. The main contribution of our proposal is the simplicity of the computational machinery required to implement the scheme, resulting in a very simple and economic hardware implementation. It is well-known that the seeming low attention towards security aspects shown by issuers is actually the right compromise of a trade-off between costs to implement radical innovations and costs to refund customers victims of fraud. This explains why the aspects related to the practical feasibility of any proposed innovation to harden the credit card transaction processing is definitely important.

References

1. Borisov, N., Goldberg, I., Wagner, D.: Intercepting mobile communications: the insecurity of 802.11. In: MobiCom 2001: Proceedings of the 7th annual international conference on Mobile computing and networking, pp. 180–189. ACM Press, New York (2001)
2. ECMA. ECMA-182: Data Interchange on 12,7 mm 48-Track Magnetic Tape Cartridges — DLT1 Format (December 1992)
3. Hill, J.R.: A table driven approach to cyclic redundancy check calculations. SIGCOMM Comput. Commun. Rev. 9(2), 40–60 (1979)
4. Li, Y., Zhang, X.: A security-enhanced one-time payment scheme for credit card. In: Proceedings of the 14th International Workshop on Research Issues on Data Engineering: Web Services for E-Commerce and E-Government Applications (RIDE 2004), pp. 40–47 (2004)
5. Li, Y., Zhang, X.: Securing credit card transactions with one-time payment scheme. Electronic Commerce Research and Applications 4, 413–426 (2005)
6. Luhn, H.P.: Computer for verifying numbers. U.S. Patent 2, 950, 048 (1960)
7. NIST/NSA. Fips 180-2 secure hash standard (SHS). NIST/NSA (August 2002)
8. Dynamic passcode authentication, http://www.visaeurope.com
9. Private Payments, http://www10.americanexpress.com
10. Paypal, http://www.paypal.com
11. Peterson, W.W.: Error-correcting codes. MIT Press and J. Wiley & Sons (1961)
12. Ramabadran, T.V., Gaitonde, S.S.: A tutorial on crc computations. IEEE Micro. 8(4), 62–75 (1988)
13. Rubin, A., Wright, N.: Off-line generation of limited-use credit card numbers. In: Proceedings of the Fifth International Conference on Financial Cryptography, pp. 165–175 (2001)
14. Sarwate, D.V.: Computation of cyclic redundancy checks via table look-up. Commun. ACM 31, 1008–1013 (1988)
15. SET Secure Electronic Transaction LLC. SET Secure Electronic Transaction Specification, http://www.setco.org
16. Shamir, A.: Secureclick: A web payment system with disposable credit card numbers. In: Syverson, P.F. (ed.) FC 2001. LNCS, vol. 2339, pp. 232–242. Springer, Heidelberg (2002)
17. Singh, A., dos Santos, A.L.M.: Grammar based off line generation of disposable credit card numbers. In: SAC 2002: Proceedings of the 2002 ACM symposium on Applied computing, pp. 221–228. ACM Press, New York (2003)
18. Singh, A., dos Santos, A.L.M.: Context free grammar for the generation of one time authentication identity. In: FLAIRS Conference (2004)
19. Stubblefield, A., Ioannidis, J., Rubin, A.D.: A key recovery attack on the 802.11b wired equivalent privacy protocol (wep). ACM Trans. Inf. Syst. Secur. 7(2), 319–332 (2004)

Online Privacy: Measuring Individuals' Concerns

Maria Moloney and Frank Bannister

Trinity College Dublin
Maria.Moloney@cs.tcd.ie, fbnnistr@tcd.ie

Abstract. Existing research within the Information Systems domain has shown that there is a substantial level of online privacy concern among the online community. However it is not clear from an extensive review of the literature that the complete set of online privacy concerns has yet been identified or whether the concerns that have been investigated, by way of surveys, have adequate theoretical foundations. This paper considers the work of two prominent privacy theorists Westin and Altman, and from their privacy theories infers a set of online privacy concerns. These inferred privacy concerns are then compared with a list of online privacy concerns drawn from the empirical literature. This comparison highlights the similarities and inconsistencies between both sets of concerns. From the findings, an online privacy model is devised which attempts to outline the components of the concept of online privacy and their interdependencies. By representing the concept of online privacy in the form of a model, areas where concern arises can be highlighted more easily and as a result measures can be taken to reduce such concern.

Keywords: Information Systems, Internet, Privacy, Trust, Data Security.

1 Introduction

In the last twelve months, there have been a number of high profile breaches of privacy reported in the media. In early 2008, a laptop with the confidential records of more than 170,000 Irish blood donors and 3,200 patients was stolen in New York. The stolen records included names, genders, dates and places of birth, telephone numbers and the blood groups of individuals who had given blood (Heffernan and Kennedy, 2008). The UK's Ministry of Defence lost three laptops containing personal details of hundreds of thousands of military recruits. The files contained names, addresses, passport details, national insurance numbers, drivers' licence details, family details, doctors' addresses, NHS numbers and some bank details of those who joined, or inquired about, the armed forces (Drury, 2008). In late 2007, Revenue and Customs in the UK lost the personal details of nearly half of the UK's population when they lost the entire child benefit database in the post (Webster et al., 2007).

These breaches only reflect Ireland and the UK. In the US, over 217 million records containing personal information have been involved in security breaches since 2005 (Privacy Rights Clearinghouse, 2005). Given the frequency and magnitude of these breaches, it is little wonder that concern is growing among the public for the safety of their personal information.

G. Psaila and R. Wagner (Eds.): EC-Web 2008, LNCS 5183, pp. 21–30, 2008.

In this paper, it is argued that an interdisciplinary approach is required to resolve the ethical issue of protecting an individual's private space. Consequently, the paper draws on the disciplines of law, social studies and information systems. The paper is organised as follows. The next section examines some theories and definitions of privacy from within these disciplines. Section three covers the formulation of the research question and section four contains a brief outline of the proposed research methodology. This is followed by a short conclusion.

2 Review of Relevant Literature

According to Westin (2003), whenever a privacy claim is recognised in law or social convention, we speak of "privacy rights". Privacy has been declared as a fundamental right for every human in many enduring bodies of law, the principle ones being Article 12 of the Universal Declaration of Human Rights (United Nations, 1948), article 8 of the European Convention on Human Rights (ECHR) (The Council of Europe, 1950), article 7 and 8 of The Charter of Fundamental Rights of the European Union (The Council of Europe, 2000).

However, to establish privacy as a fundamental human right is futile if the very notion of privacy is not understood. This paper uses privacy definitions from two domains that have contributed greatly towards the formulation of the privacy concept to which our modern society adheres. These domains are the ethical and legal domains.

A challenge, and an opportunity, in this research is that there is no agreed definition of what constitutes privacy. To give a flavour of the complexity of the problem, consider the following approaches to the subject. The theorist Charles Fried (1990) believes that privacy is not simply an absence of information about a person in the minds of others, rather it is the control that a person has over information about themselves. This view is somewhat incomplete in that privacy can be interpreted as the control that a person has over information about themselves that they wish to keep from others. As a reaction to Fried's 'control theory' of privacy Moor (1997) proposes a 'restricted access' view of privacy. Rather than regarding privacy as an all or nothing proposition, Moor regards it as a complex of situations in which information is authorized to flow to specific people, at specific times. He argues that in a highly computerized culture, it is simply impossible to control all personal information that resides on computer systems around the world. Therefore, the best way to protect our privacy is to make sure the right people have access to relevant information at the right time. Moor calls this view of privacy, the restricted access view, which has the added advantage of Fried's control theory in that it gives individuals as much control over personal data as realistically possible. For this reason he labels his theory the "control/restricted access" theory of privacy (Moor, 1997).

An advocate of Moor's control/restricted access theory is Herman Taviani. Taviani also points out that modern privacy theorists tend to analyse the notion of privacy in terms of controlling the flow of personal information and have coined the phrase "informational privacy" to express this new concept (Taviani, 2007). The term, informational privacy, is often used when referring to an individual's online privacy

(Although it should be noted that, even though protecting an individual's personal information once submitted online is a key area for concern, submitting personal information is not an individual's sole reason for spending time online. Both social and non-social activities make up for a large portion of an individual's time spent online (Zhao, 2006)).

The variety of views can be further illustrated by looking at the work of arguably the two key figures in the privacy domain of the last thirty years, namely Alan Westin and Irwin Altman. Both put forth their own theories of privacy in the 1970's and have since enjoyed considerable recognition from their peers (Boritz et al., 2006, Goodwin, 1992, Margulis, 2003a, Margulis, 2003b, Margulis et al., 2006, Marshall, 1974, Nissenbaum, 1998, Parent, 1983, Rotenburg, 2000, Taviani, 2007).

Westin (1970) defines privacy as:

"the claim of individuals, groups or institutions to determine for themselves when, how, and to what extent information about them is communicated to others"

(Westin, 1970, Section One). He delineates four states of privacy:

- Solitude is the state of being free from the observation of others;
- Intimacy is the state that in a small group, seclusion fosters close relationships;
- Anonymity is freedom from identification and from surveillance in public places;
- Reserve is based on limiting disclosure to others and requires others to recognise and respect that desire.

  He also outlines four functions or reasons for privacy:

- Personal autonomy is the desire to avoid being manipulated, dominated, or exposed by others;
- Emotional release is the release of tensions under social restrictions like role demands, emotional states or minor deviances;
- Self-evaluation deals with extracting meaning from personal experiences and exerting individuality on events;
- Limited communication sets interpersonal boundaries and protected communication allows for sharing personal information with trusted others (Westin, 2003).

In contrast, Altman (1975) defined privacy simply as:

"the selective control of access to the self".

He sees 'selective control' as intrinsic to privacy because people try to control their openness or distance to others by being open and available or closed and unavailable at different times and for different reasons (Altman and Chemers, 1980). Altman argues that, privacy is a dynamic process whereby people vary in the degree to which they make themselves accessible to others. He describes the crucial idea of his framework as being that privacy is a central concept that provides a bridge between personal space, territory and other realms of social behaviour. He describes privacy as *"an interpersonal boundary regulation process by which a person or group regulates interaction with others"* (Altman and Chemers, 1980, p75).

This research uses a combination of Westin and Altman's privacy theories as the basis for developing an online privacy model. The next section outlines in detail the process behind the creation of the model.

3 Formulation of the Research Question

In arriving at the research question, the theoretical foundations of privacy were first explored by examining in depth both Westin and Altman's theories of privacy. Individual sets of potential privacy concerns were then derived from both theories. Table 1 summarises both Westin (1970) and Altman's (1975) theories of privacy and lists some potential privacy concerns that could arise from a denial of privacy.

Table 1. Inferred privacy concerns derived from Westin and Altman's Privacy Theories

Westin's Privacy Functions	Potential Privacy Concerns
Personal autonomy, Emotional release, Self-evaluation, Limited & protected communication.	If privacy is denied an individual they may feel: exposed and open to the threat of manipulation or domination. unable to express emotional release as a result of being denied the ability to release tensions that have built up in various ways under certain social conditions. unable to extract personal meaning from events and an inability to react to those events in a personal way. unable to set interpersonal boundaries and share information with other trusted individuals.
Altman's Privacy Theory	**Potential Privacy Concerns**
To allow for: The variation in degrees of openness or distance, in response to changes in internal states and external conditions. The movement from the actual to the desired level of privacy. A change in the desired level of privacy by increasing and decreasing the level from time to time. The bi-directional movement of inputs from others and outputs to others.	If privacy is denied an individual they may feel: unable to regulate the access of others to themselves. unable to move from the actual state to the desired state of privacy. unable to regulate the bi-directional movement of inputs from others and outputs to others.

A key distinction between these two privacy theories is that Westin interprets privacy as *a series of personal actions taken by the individual in order to function comfortably in their environment.* Altman interprets privacy as *a continuous reaction by the individual to both internal and external changes,* which causes the individual to engage in a continual process of adjustment from their actual level of privacy to their desired level of privacy.

Having explored the theoretical foundations of privacy, the second step was to explore the concept of privacy in a modern setting and to examine whether these theoretical foundations, developed over thirty years ago, still apply today. The modern setting was online and the concerns of individuals regarding their privacy were examined by looking at evidence from the empirical literature.

Like the traditional concept of privacy, online privacy is defined in various ways across different disciplines. This paper uses Westin's (2003) definition of online privacy (which is essentially an extension of the definition cited above):

"the claim of an individual to determine what information about himself or herself should be known to others...This also involves when such information will be obtained and what uses will be made of it by others"

(Westin, 2003, P431). However a number of academic studies have used surveys to assess the level of privacy concern among the Internet community and the findings of these do not sit well with Westin's definition. Examples of recent academic surveys of privacy are Ackerman et al.(1999), Agre and Rotenberg (1997), Dhillon and Moores (2001), Harris Interactive (2002), Milberg et al. (1995), Smith et al. (1996). Table 2 below summarises the online privacy concerns that were highlighted in these six surveys.

Table 2. Summary of Information Privacy Findings

Information Privacy Concerns	Researcher(s)
Collection of personal information	Smith et al. (1996) Milberg et al. (1995) Ackerman et al. (1999) Harris Interactive (2002)
Unauthorized secondary use of personal information	Smith et al. (1996) Dhillon and Moores (2001) Milberg et al. (1995) Ackerman et al. (1999) Harris Interactive (2002),
Improper access to personal information	Smith et al. (1996) Dhillon and Moores (2001) Milberg et al. (1995) Harris Interactive (2002)
Errors with stored personal information	Smith et al. (1996) Milberg et al. (1995)
Identification theft through personal information once it has been collected;	Dhillon and Moores (2001) Ackerman et al. (1999)
Receiving unsolicited email.	Dhillon and Moores (2001)
Monetary incentives decrease reluctance to impart personal information	Hui et al. (2007)
The more information requested, the less likely the subjects were to disclose it	Hui et al. (2007)
The purpose for collecting information.	Ackerman et al. (1999)

A number of academic studies have also been used to ascertain the antecedents to online privacy concerns (Ackerman et al., 1999, Castañeda and Montoro, 2007, Dinev and Hart, 2004, George, 2004, Harris Interactive, 2002, Milne and Boza, 1999, Phelps et al., 2000, Phelps et al., 2001). An understanding of antecedents is critical to the development of effective online privacy policies and practices (Phelps et al., 2001). Table 3 below shows a summary of the most common antecedents derived from eight surveys examined by this research.

Table 3. Antecedents to Online Privacy Concerns

Antecedents to Online Privacy Concerns	Researcher(s)
Lack of trust in a company's ability to adequately safeguard personal information	(Ackerman et al., 1999, Dinev and Hart, 2004, George, 2004, Harris Interactive, 2002)
The lack of knowledge regarding a company's use of such information including collection and redistribution	(Ackerman et al., 1999, Castañeda and Montoro, 2007, Dinev and Hart, 2004, Harris Interactive, 2002, Phelps et al., 2001)
The sensitivity of the information that is being requested	(Castañeda and Montoro, 2007, Milne and Boza, 1999, Phelps et al., 2000)
The lack of informational control individuals feel once they have imparted their personal information.	(Castañeda and Montoro, 2007, Milne and Boza, 1999, Phelps et al., 2000)

From the information outlined in tables 1,2 and 3, a remarkable difference between theory and practice regarding privacy concerns came to light; While the concept of trust in relation to privacy features heavily among the online privacy concerns, Westin and Altman fail to mention it directly. Westin (1970) touches upon it when he describes the need for privacy to achieve personal autonomy. He describes individuals as having different layers of protection around themselves and only a few close individuals are allowed to enter into the innermost layers closest to the true self. Similarly, Altman and Taylor (1973) use an onion metaphor in their social penetration theory to explain the same phenomenon. An individual's personality has multiple layers, like an onion. As people get to know one another they penetrate each layer to come closer to the core of the individual. However, neither theorist incorporates the notion of trust into their privacy theories.

The concern about relinquishing control over personal space was evident in both theory and practice. Such concern indicates that individuals need to feel in control of their privacy at all times both on and offline. Furthermore, it is clear that individual's privacy concerns are based on both real risks (such as identity theft) and a personal desire for privacy, even where no real risks exist. Trust enables the individual to feel more in control of a situation when the risks of the situation are unknown. Essentially, the degree of trust that individuals have in a third party affects their perception of the level of risk involved in any transaction with that third party.

Drawing together these ideas, the following preliminary questions need to be explored in more depth before the main research questions can be presented:

- Is trust is a prerequisite for individuals to impart personal information online?
- Do individuals need to feel in control of their personal space while online?
- Should a privacy theory incorporate both concepts of trust and control in some manner if it is to be relevant in an online environment?

The next two sub-sections discuss these questions in light of the existing literature on online trust and Rotter's (1966) locus of control formulation and derives the final research question.

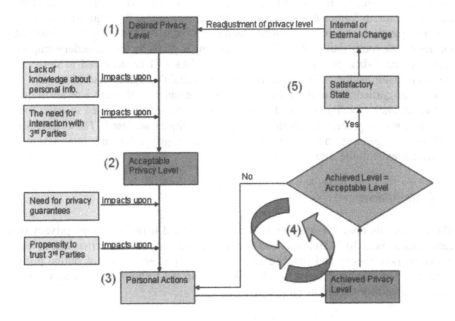

Fig. 1. A Representation of Privacy

3.1 The Main Research Questions

This representation of privacy is based on evidence from the existing literature, both theoretical and empirical. However, given the inconsistencies between theory and practice highlighted earlier in the research, it is not clear whether this is a true representation of privacy. Consequently, the findings need to be validated using primary data before they can be deemed rigorous. The objective of this research is to establish the validity of the above model and address the following related questions:

1. What are the online privacy concerns of individuals? Privacy concerns will arise when the actual level of privacy falls short of the acceptable level of privacy.
2. What are the antecedents to these privacy concerns? In particular, what is the role of control and trust in these?
3. Once these have been established how can they:
    A. Be represented in a model?
    B. Be addressed/reduced?

4 Outline of Research Design and Implementation

At the time of writing, the research methodology has yet to be finalised. This section outlines the approach currently proposed. This research will combine both qualitative and quantitative research approaches. The qualitative approach will take the form of focus groups and the quantitative approach will be in the form of an online questionnaire. The questionnaire will be drawn from existing instruments cited above, but modified to reflect the research question. The findings of earlier surveys will also be used as a guide to formulate questions for the focus groups and the questionnaire.

Open questions will be used during the focus groups and the answers will be audio recorded. The focus groups will discuss issues highlighted by the secondary empirical and theoretical data previously examined. The data will be analysed to detect key themes or patterns that can be examined in more detail through the questionnaire.

The data gathered from the questionnaire will be analysed through a series of quantitative analysis techniques. The data will first be sorted and analysed in order to ensure the most appropriate statistical analysis techniques are used. These techniques will aid in exploring and describing the data and highlighting any relationships or trends that exist in the data.

5 Conclusion

The goal of this research is to determine the extent and nature of online privacy concerns experienced by individuals and the antecedents to these concerns. The research draws on both theory and practice to derive a comprehensive set of online privacy concerns. The results of the research will facilitate the formulation of a new privacy model adapted for the online world.

References

1. Ackerman, M.S., Cranor, L.F., Reagle, J.: Privacy in E-Commerce: Examining User Scenarios and Privacy Preferences. In: The 1st ACM conference on Electronic commerce Denver, Colorado, United States. ACM Press, New York (1999)
2. Agre, P.E., Rotenberg., M.: Technology and Privacy: The New Landscape. MIT Press, London (1997)
3. Altman, I.: The Environment and Social Behavior, Monteray, California. Brooks/Cole (1975)
4. Altman, I., Chemers, M.: Culture and Environment, Monteray, California. Brooks/Cole Publishing Company (1980)
5. Altman, I., Taylor, D.: Social Penetration: The Development and Dissolution of Interpersonal Relationships, London, Holt. Rinehart and Winston (1973)
6. Ambrose, P.J., Johnson, G.L.: A Trust Model of Buying Behavior in Electronic Retailing. The Association for Information Systems, Baltimore (1998)
7. Barber, B.: The Logic and Limits of Trust, New Brunswick, New Jersey. Rutgers University Press (1983)
8. Boritz, E., No, W.G., Sundarraj, R.P.: Internet Privacy: Framework, Review and Opportunities for Future Research (2006), http://ssrn.com/abstract=908647
9. Castañeda, J.A., Montoro, F.J.: The effect of Internet general privacy concern on customer behavior. Electronic Commerce Research 7, 117–141 (2007)
10. Covello, V.T.: Trust and Credibility in Risk Communication. Health Environment Digest 6, 1–4 (1992)
11. Dhillon, G.S., Moores, T.T.: Internet Privacy: Interpreting Key Issues. Information Resources Management Journal 14, 33–37 (2001)

12. Dinev, T., Hart, P.: Internet Privacy Concerns and their Antecedents - Measurement Validity and a Regression Model. Behaviour and Information Technology 23, 413–422 (2004)
13. Drury, I.: Oops! MoD lost THREE laptops, not just one, admits minister. The Daily Mail (2008),
    http://www.dailymail.co.uk/pages/live/articles/news/news.htm
    l?in_article_id=509629&in_page_id=1770&in_page_id=1770&expan
    d=true ed
14. Fried, C.: Privacy: A Rational Context. In: Erman, M.D., Williams, M.B., Gutierrez, C. (eds.) Computers, Ethics & Society, New York. Oxford University Press, Oxford (1990)
15. Gefen, D.: E-Commerce: The role of familiarity and trust. The International Journal of Management Science 28, 725–737 (2000)
16. George, J.F.: The Theory of Planned Behavior and Internet Purchasing Internet Research: Electronic Networking Applications and Policy, vol. 14, pp. 198–212 (2004)
17. Gershaw, D.: Line on Life: AIDS and Teenagers (1989),
    http://virgil.azwestern.edu/~dag/lol/AIDSTeens.html
18. Goodwin, C.: A Conceptualization of Motives to Seek Privacy for Nondeviant Consumption. Journal of Consumer Psychology 1, 261–284 (1992)
19. Interactive, H.: Privacy On and Off the Internet: What Consumers Want. Harris Interactive Inc., Hackensack, New Jersey, Privacy & American Business (2002)
20. Heffernan, B., Kennedy, E.: Alert as 170,000 blood donor files are stolen The Irish Independant, Dublin (2008), http://www.independent.ie/national-news/
    alert-as-170000-blood-donor-files-are-stolen-1294079.html
21. Hui, K.L., Teo, H.H., Lee, S.Y.T.: The Value of Privacy Assurance: An Exploratory Field Experiment. MIS Quarterly 31, 19–33 (2007)
22. Kim, K.K., Prabhakar, B.: Initial Trust and the Adoption of B2C e-Commerce: The Case of Internet Banking. Database for Advances in Information Systems Control Journal 35, 50–65 (2004)
23. Lee, M.K.O., Turban, E.: A Trust Model for Consumer Internet Shopping. International Journal of Electronic Commerce 6, 75–91 (2001)
24. Levenson, H.: Multidimensional Locus of Control in Psychiatric Patients. Journal of Consulting and Clinical Psychology 41, 397–404 (1973)
25. Liu, C., Marchewka, J.T., Lu, J., Yu, C.-S.: Beyond Concern - A Privacy-Trust-Behavioral Intention Model of Electronic Commerce. Information & Management 42, 289–304 (2005)
26. Margulis, S.T.: On the Status and Contribution of Westin's and Altman's Theories of Privacy. The Journal of Social Issues 59, 411–429 (2003a)
27. Margulis, S.T.: Privacy as a Social Issue and Behavioral Concept. The Journal of Social Issues 59, 243–262 (2003)
28. Margulis, S.T., Pope, J.A., Lowen, A.: The Harris-Westin's Index of General Concern About Privacy: An Attempted Conceptual Replication, Seidman College of Business. Grand Valley State University, Grand Rapids, Michigan (2006)
29. Notes
30. Marshall, N.: Dimensions of Privacy Preferences. Multivariate Behavioral. Research 9, 255–271 (1974)
31. Mcknight, D.H., Chervany, N.L.: What Trust Means in E-Commerce Customer Relationships: An Interdisciplinary Conceptual Typology. International Journal of Electronic Commerce 6, 35–59 (2001)
32. Mcknight, D.H., Cummings, L.L., Chervany, N.L.: Initial Trust Formation in New Organisational Relationships. Academy of Management Review 23, 473–490 (1998)
33. Milberg, S.J., Burke, S.J., Smith, H.J., Kallman, E.A.: Values, Personal Information Privacy, and Regulatory Approaches. Communications of the ACM 38, 65–74 (1995)
34. Milne, G.R., Boza, M.E.: Trust and Concern in Consumers' Perceptions of Marketing Information Management Practices. Journal of Interactive Marketing 13, 5–24 (1999)

35. Moor, J.H.: Towards a theory of privacy in the information age. ACM SIGCAS Computers and Society 27, 27–32 (1997)
36. Nissenbaum, H.: Protecting Privacy in an Information Age: The Problem of Privacy in Public. Law and Philosophy 17, 559–596 (1998)
37. Parent, W.A.: A New definition of Privacy for the Law. Law and Philosophy 2, 305–338 (1983)
38. Phelps, J., Nowak, G., Ferrell, E.: Privacy Concerns and Consumer Willingness to Provide Personal Information. Journal of Public Policy & Marketing 19, 27–42 (2000)
39. Phelps, J.E., D'souza, G., Nowak, G.J.: Antecedents and Consequences of Consumer Privacy Concerns: An Empirical Investigation. Journal of Interactive Marketing 15, 2–17 (2001)
40. Privacy Rights Clearinghouse, A Chronology of Data Breaches. Privacy Rights Clearinghouse, San Diego (2005)
41. Rotenburg, M.: What Larry Doesn't Get: Fair Information Practices and the Architecture of Privacy. Stanford Technology Law Review (2000)
42. Rotter, J.B.: Generalized expectancies for internal versus external control of reinforcements. Psychological Monographs 80, 1–28 (1966)
43. Shankar, V., Urban, G.L., Sultan, F.: Online Trust: A Stakeholder Perspective, concepts, implications and Future Directions. Journal of Strategic information Systems 11, 325–344 (2002)
44. Smith, H.J., Milberg, S.J., Burke, S.J.: Information Privacy: Measuring Individuals' Concerns about Organizational Practices. MIS Quarterly 20, 167–196 (1996)
45. Taviani, H.T.: Philosophical Theories of Privacy: Implications for An Adequate Online Privacy Policy. Metaphysics 38 (2007)
46. The Council of Europe, The European Convention on Human Rights. C364 (1950)
47. The Council of Europe, The Charter of Fundamental Rights of the European Union. C364 (2000)
48. United Nations, Universal Declaration of Human Rights. General Assembly resolution (1948)
49. Webster, P., O'Neill, S., Blakely, R.: 25 million exposed to risk of ID fraud. Times Online, London (2007),
http://www.timesonline.co.uk/tol/news/uk/article2910705.ece
50. Westin, A.F.: Privacy and Freedom. The Bodley Head Ltd., London (1970)
51. Westin, A.F.: Social and Political Dimensions of Privacy. The Journal of Social Issues 59, 431–453 (2003)
52. Zhao, S.: Do Internet Users have more Social Ties? A Call for Differentiated Analyses of Internet Journal of Computer-Mediated Communication 11 (2006)

Analysing the Key Factors of Web Design: A Heuristic Evaluation

Carlos Flavián, Raquel Gurrea, and Carlos Orús

University of Zaragoza, Department of Economy & Business Studies, Gran Vía 2,
50005 Zaragoza, Spain
cflavian@unizar.es, gurrea@unizar.es, corus@unizar.es

Abstract. Due to the rapid growth of the electronic commerce and consequently the increase of the competence, a special interest among the specialized
literature is arising focusing on the critical factors which could lead to the e-
commerce success. In this sense, the web design has been revealed as a critical
aspect in order to achieve positive outcomes of the users. Consequently, this research is focused on analyzing the importance of web design for the success of
a virtual store. A heuristic evaluation has been developed in order to identify
the good and bad design practices of three well-known websites. This method is
particularly useful for our study, since the opinions from multidisciplinary experts give us a more complete understanding about what is the most important
for the design of the websites. The results of the test, together with a review of
the specialized literature, allowed us to offer some managerial implications and
recommendations for designers in order to carry out a successful virtual store.

Keywords: web design, usability, atmospherics, electronic commerce success,
heuristic.

1 Introduction

In the last years, the diffusion of Internet as a new retail channel is being developed
with a great growth. Sales through the web are up to 70 billon US$ in United States in
the second half of 2007, which means an increase of 20% for the same period last
year [1]. For Europe, the total sales of B2C e-commerce reached 133 billion US$ in
2007, and the forecasts establish a sales outcome of 407 billion US$ by 2011 [2]. In
this market of increasing competence, with a target of more than 1300 million people
[3] and more than 70 thousand million dot-com websites in the world [4], there has
been arising a great body of research focused on the factors that affect the success of
an e-commerce website. In this line, many studies have identified the web design as a
key factor for the development of a good interface for satisfying the consumer needs.
A good design is relevant for companies to survive in the extremely competitive
World Wide Web (e.g. [5], [6], [7]). Moreover, web design is important for getting
higher levels of satisfaction with the website [8] or for increasing the online purchase
intention of the consumer [9]. Thus, the goal of this paper is to identify the key factors
that could determine the success of the websites, emphasizing the specific aspects
related to web design. With this aim, we develop a heuristic evaluation in order to

G. Psaila and R. Wagner (Eds.): EC-Web 2008, LNCS 5183, pp. 31–40, 2008.
© Springer-Verlag Berlin Heidelberg 2008

know good and bad practices carried out by recognized e-businesses. Finally, we propose some conclusions and managerial guidelines for the achievement of adequate structures and design of the websites.

2 The Relevance of Website Design for e-Commerce Success

A great number of authors have made efforts in order to delimit the concept associated to a successful website. In general terms, a successful website, in the context of the electronic commerce, "is one that attracts customers, makes them feel the site is trustworthy, dependable, and reliable and generates customer satisfaction" [10]. In this line, there is a growing interest for the main factors that affect the levels of acceptance and success of the website. We have to point out the research made by Jarvenpaa and Tood [11], who examined online consumers' perceptions with the aim of identifying the aspects that could influence on their attitudes and intentions to buy on the Internet. These authors affirmed that the aspects related to the perceptions about the products, the shopping experience (e.g. required effort, convenience), the services (e.g. adequate information and attractive appearance) and the perceived risk (e.g. pay, possible dissatisfaction), affect significantly the consumers' attitudes and intentions to buy. In the same line, Lohse and Spiller [12] analyzed the relationship between the features of the virtual store and the traffic and sales levels on the website. The results suggested that the costs associated to the information search processed are the main determinant of those dependent variables. In this way, the authors noted the importance of the website design and more specifically, the easy of navigation for increasing the number of visits to a website. Thus, we could point out the aspects related to the convenience of the new digital medium, which is one of the main advantages of the electronic commerce for the consumers [13], [14]. The Internet users also value the importance of the security and privacy of the transactions, as it has been recognized in the specialized literature (e.g. [15], [16]).

In the same way, we could note the relevance of presenting high quality information, good contents and an efficient and attractive navigation. These aspects have been identified by [17] and [18] as the most important advantages of the online activity. Moreover, it seems remarkable the relevance of the website design for getting optimal results in the virtual channel. Following this idea, we could affirm that the design of websites has been largely studied from multiple perspectives, most of them have identified the factors that could influence on website degree of acceptation and success (e.g. [19], [20], [21], [22], [23]).

In this line, the usability studies what elements a website must have so the consumer can manage it easily. These aspects could lead to higher levels of satisfaction, trust and loyalty towards a specific website [8], [24], [25]. Nielsen [26] defines the usability of a website as the ease with which the user can learn to manage the system and memorise the basic functions, the efficiency of design of the site, the degree of error avoidance and the general satisfaction of the user. More specifically, the usability is a quality attribute that assesses how easy user interfaces are to use, and we can identify five dimensions or quality attributes: learnability, efficiency, memorability, errors, and satisfaction [27]. Following this line, the usability can also be understood as a tool for measuring the quality of a website [5]. Thus, there have been studies indicating that usability improves the best understanding of the contents and tasks that the consumer has to know for the achievement of a goal, which in turn reduces the probability of error and improves the level of trust [28]. Besides, usability is related to the consumer ability to identify where he or she is and what he or she can do in each moment of the navigation. A good design should ensure a high level of usability [29]. An attractive design can evoke feelings of pleasure in the use of a website. As a consequence, the best usability, in the sense of a comfortable atmosphere, can create a positive bias in the consumer.

It is remarkable the study carried out by Palmer [30], who proposed and validated measures of the websites' usability and design, identifying aspects related to the response time (download delay), the organization of the contents (navigation), and the information and contents of the website (content). The research also established that the characteristics of the media richness, such as the capability of the website for customizing the appearance and the contents (interactivity), and the presence of feedback between the vendor and the consumer, were determinants of the success of a website.

In a similar way, the marketing literature has studied how the factors that define the store environment influence the consumer's mood and purchase behaviour [31], [32], [33], [34]. Design factors related to visual cues (e.g., layout, colour) are important for getting a positive response of the consumer and to facilitate consumer goals [35]. This issue has been taken to the web environment, and has been adapted to e-commerce. Dailey [36] defines web atmospherics as the intended design of web environments to generate positive effects (cognitive and affective) on the consumer in order to increase positively the consumer responses. Among these responses we can find a major number of visits or a major time browsing in the website, in order to turn the user of a website into a client. We can mention studies as of Dailey [37] and Eroglu et al., [35] who analyzed the influence of web atmosphere on consumers, and demonstrated that the insights of this atmosphere influence on the cognitive and affective states of the consumer, and as a consequence on the purchase behaviour towards the website.

3 Methodology

Once we have identified the key aspects that are critical for achieving a good design and therefore a successfully e-commerce website, in the following section we carry out a heuristic evaluation in order to know how well or bad the e-businesses are developing their activities through the Internet, with the final goal of establishing some guidelines that may help the designers to implement the best choices in their websites and getting higher levels of success.

The heuristic evaluation [38], [39] is a method for finding the design problems and good practices in a user interface. This evaluation involves having a small set of evaluators who examine the interface and judge its compliance with recognized usability principles, namely "heuristics". We selected a group of academic and business experts on both formal and conceptual issues. The experts' group was selected in accordance with their professional experience and knowledge of the context from both the academic and the professional perspectives. Specifically, the group was multidisciplinary as it included five experts in different areas such as Marketing, Usability, Information Systems and New Technologies.

During the evaluation session, the evaluator goes through the interface several times and inspects the various dialogue elements and compares them with a list of recognized heuristic principles. These heuristics are general rules that seem to describe common properties of usable and well-design interfaces. In this way, we developed a heuristic test according to Sutcliffe [40], who included measures of the appearance of the websites as well as the usability principles -heuristics- established by [39], and purposed an assessment process based on three stages concerning the website itself. Firstly, an evaluation of the attractiveness of the website is made in order to check out if the website is able to attract the users' attention. Secondly, the aspects related to navigation and usability of the website are assessed, where the structure and the contents play an important role. Finally, the design of the website must be focused on provide an efficient transaction process, so evaluators score different characteristics such as the transaction paths, feedback or security.

Thus, we focus this research on the practices of some of the most recognized websites in the last time, attending to several references in research articles (e.g., [41], [42], [43], [44], [45], [46]), awards and rankings of the best designed websites (e.g. [47], [48], [49]). In this line, the test was applied to the Amazon (www.amazon.com) and Apple store (www.store.apple.com) interfaces.

4 Heuristic Evaluation

Regarding the visual appearance of the front page of Amazon (http://www.amazon.com/), experts' evaluations state that the website presents a clear presentation, using non-saturated background colour, which avoids an overload in the users' mind. The main menu and the search engine are highlighted in different blue colours on the left-side and on the top-centre of the site respectively. According to the experts, this allows Amazon to attract the user's attention, since the structure of the menus and the search engine of this website is one of its greatest strengths [50], [51]. The central part of the website contains the product novelties, offers and sales, with a use of good quality images and colourful headlines, which underlines the updating feature of the site and reinforces its appearance. The scroll of the front page is pointed as a bit long, which may cause a negative effect on the user (in a sense of too much information in the home page) or may be an unnecessary effort made by the website designers (the user may not bother scrolling down to the bottom of the page and does not watch this part).

The main source of success of Amazon is based on the features related to its structure and navigation [35], which has been supported by the experts' reports. In this sense, we could stress the quality of the search functions (http://www.amazon.com/gp/site-directory/ref=_gw_; http://www.amazon.com/audio-video-portable-accessories/b/ref= sd_allcat_av?ie=UTF8&node=1065836). All the experts agreed about the simplicity of the navigation and the fitness of the structure of the contents, which implies an adequate degree of users' freedom within the website. The contents about the products are organized in a logical way, where each category presents subcategories and specific items. Besides, the evaluators pointed out the good quality of the specific information about products, qualifying it as "complete", "comprehensive", or "accuracy", as well as the availability of additional information (i.e. reviews, good quality of images, 360° views). However, the evaluators noted the absence of a "map of the site" button in order for users to control the navigation. In this line, we could advise to make shorter scroll down cutting out the content of each page or by means of the use of extending labels.

In the sense of the shopping or transaction process, the five evaluators complaint about the fact of the users have to register in the Amazon's website in order to make a purchase, even when the user only wants to know how the site performs the shopping orders. Thus, it would be desirable to offer the possibility of making a simulated purchase or to explain how the shopping process is carried out. In spite of that, the experts' evaluations for the questions related to the transactions are quite positive. The shopping process is completed is four simple stages. In fact, one of the most salient reasons for the Amazon's success is its one-click-to purchase [12], where the registered users can make a purchase with just one click a stage. Besides, the evaluators positively assessed the existence of relevant information related to security, privacy and delivery issues.

With reference to the visual aesthetics of the Apple Store website, the experts coincide to qualify a smart appearance, where the use of soft colours gives the impression of elegancy and sobriety (http://store.apple.com/us), and it also enhances the beauty of its products. The front page of the website is exclusively dedicated to the products (there is no banners, flashes or links to other sites non-related to Apple), and the product categories and search function are displayed in the left-side of the page as well as at the top. All this stresses the importance that Apple gives to its products, so that the goal could be to focus the users' attention only on the products offered by the company. The evaluations also note the presence of updated information and an adequate scroll down.

The structure of the contents and the navigation are also positively valued by the experts. The experts noted the simple navigation, characterized by straight routes through the product categories. With reference to the information related to the products, the experts stress two main aspects: first, the short scroll allows users have an overview of all the relevant information about the product. Besides, there is a little bar below the product itself which leads the user to a great amount of appropriate information relevant for the purchase decision (http://store.apple.com/us/browse/home/ shop_ipod/family/ipod_nano?mco=NzY2NzYz). The location of that bar was identified as a design problem, since it may go unnoticed by users; second, the simplicity of the design permits the use of high-quality images of the products, accessories and so on. Again, the experts noted the lack of a map of the site.

Regarding the design aspects related to the shopping process, the overall evaluation is also favourable. The experts appreciated the visual transaction path, made in four stages and a one-click shopping -as well as Amazon website- (http://store.apple.com/ us/browse/home/shop_ipod/family/ipod_nano?mco=NzY2NzYz). The presence of a barrier was also identified for the next step, since registration is needed, so that the user can't keep behaving as a browser. Saving this hitch, the process is assessed as easy to manage, and the evaluators find the security aspects in an efficient place, just before verifying the purchase. The results of the test also showed a good valuation of the presence of a bar with helpful tasks, and a phone number for assistance.

5 Discussion

With the aim of finding out the key factors that affect the levels of acceptation and success of the commercial websites, this research has focused on the aspects related to the website design that could influence the users and consumers' perceptions and behaviours. Specifically, a heuristic evaluation has been carried out for three recognized and high impact websites. This methodology is considered as an excellent tool for identifying design problems and for discussing the positive aspects of the web design.

Thus, the literature review and the heuristic evaluation allow us to state the importance of web design in order to get positive outcomes in the electronic commerce context. However, to achieve an efficient design is not a easy task, since there are a lot of factors to take into account, not only in terms of design itself, but also in terms of possible costs derived from the implementation and maintenance of the website. We have also to consider the possible losses in the download speed because of the use of design tools. Herein, it emerges a challenge for e-businesses, since the limited dimensions of the computer screen provoke that the designers have to carefully manage the design alternatives, displaying the options that reflect the essence of the website in the best way and satisfy the users and consumers' needs.

Bearing in mind all above, we could offer some recommendations which may found in a right way the effectiveness of a website. Firstly, given the fact that the shopping window is the same as the computer screen in an e-store [35], it seems reasonable to put our attention into the websites' appearance, since the first impression of a website determines the user evaluation of that website [52]. In this sense, the aesthetic appearance of a website could be considered as an assessing method of the website credibility [53], wherein a bad designed website may represent a good reason not to shop on it [11]. As a consequence, the development of a website with a good use of images, graphics, icons, animations or colours, may represent a potential source to offer a more vivid website and to get a positive response by the consumer [23], [54], [55]. Furthermore, the websites should provide good levels related to the download speed of the page [12]. In this sense, a more downloading pages time, derived from overloads in the contents or even in the use of images, could lead the users to avoid the website and leave it. Thus, it is necessary to get a well-balanced equilibrium between the appearance of the website and its downloading speed.

With reference to the aspects related to the navigation of the website, we could stressed the importance of a navigation characterized by simplicity that allows to users a certain degree of freedom [41], [56], which could enhance consumer's satisfaction with the website and purchase intentions [57], [58]. Besides, the user should be able to control in which place and moment he or she is during the navigation. Thus, the availability of a map of the site or a backward button is highly valued by the users. Moreover, designers have to provide sophisticated search engines, offering timely and accurate answers to the consumers' requests, since this function improves the users' valuations of a website [6].

Organizing and managing the content displayed in a website could become another key issue in order to achieve the success of an online business. Because of the fact that searching for commercial information is one of the most performed activities carried out through the Internet [59], the quality of the information provided by the website is an important factor to get higher levels of users' satisfaction [21], [60], involvement and purchase intention [58]. Consequently, it seems recommendable to display the contents with updated, comprehensible and relevant information. Again, the visual aspects may play an important role. The use of product images with a proper size or quality turns into a key question for the user could acquire a better knowledge about the product [55] and could make a more efficient shopping decision.

Finally, the characteristics of the shopping process have been revealed as a critical aspect for the achievement of a successfully e-commerce website [42], emphasizing the efficiency of a process step-by-step in a clear and easy way [35], [41]. Moreover, the designers should take care of the information quality related to the products and services supplied by the e-store [61], offering any additional information that could be useful for the consumer [12]. The importance of the security and privacy concerns over the commercial transactions has to be also considered (e.g. [5], [62]).

References

1. U.S. Census Bureau,
   http://www.census.gov/mrts/www/data/html/07Q4.html
2. Emarketer,
   http://www.emarketer.com/Report.aspx?code=emarketer_2000426
3. Internet World Stats, http://www.internetworldstats.com
4. Domain tools, http://www.domaintools.com/internet-statistics/
5. Ranganathan, C., Ganapathy, S.: Key Dimensions of B2C Web Sites. Information and Management 39, 457–465 (2002)
6. Liang, T.P., Lai, H.J.: Effect of Store Design on Consumer Purchases: Van Empirical Study of On-line Bookstores. Information & Management 39(6), 431–444 (2002)
7. Tan, G.W., Wei, K.K.: An Empirical Study of Web Browsing Behaviour: Towards an Effective Website Design. Electronic Commerce Research and Applications 5, 261–271 (2006)
8. Kim, E.B., Eom, S.B.: Designing Effective Cyber Store User Interface. Industrial Management and Data Systems 102(5), 241–251 (2002)
9. Swaminathan, V., Lepkowska-White, E., Rao, B.P.: Browsers or Buyers in Cyberspace? An Investigation of Factors Influencing Electronic Exchange. Journal of Computer-Mediated Communication (December 5, 1999), http://www.ascusc.org/jcmc/vol5/issue2/swaminathan.htm
10. Liu, C., Arnett, K.: Exploring the Factors Associated with Web Site Success in the Context of Electronic Commerce. Information and Management 38, 23–33 (2000)
11. Jarvenpaa, S.L., Todd, P.A.: Consumer Reactions to Electronic Shopping on the World Wide Web. International Journal of Electronic Commerce 1(2), 59–88 (1997)
12. Lohse, G.L., Spiller, P.: Internet Retail Store Design: How the User Interface Influences Traffic and Sales. Journal of Computer-Mediated Communication 5(2) (1999), http://jmcm.indiana.edu/vol5/issue2/lohse.htm
13. Frazier, G.L.: Organizing and Managing Channels of Distribution. Journal of the Academy of Marketing Science 27(2), 226–240 (1999)
14. Görsch, D.: Do hybrid retailers benefit from the coordination of electronic and physical channels? In: Proceeding of the 9th European Conference on Information Systems, Bled, Slovenia (2001), http://ecis2001.fov.uni-mb.si/doctoral/Students/ECIS-DC_Goersch.pdf

15. Lynch, J., Ariely, D.: Wine Online: Search Costs Affect Competition on Price, Quality and Distribution. Marketing Science 19(1), 83–103 (2000)
16. Daniel, E., Storey, C.: Online Banking: Strategic and Management Challenges. Long Range Planning 30(6), 890–898 (1997)
17. Alba, J., Lynch, J., Weitz, B., Janiszawski, C., Lutz, R., Sawyer, A., Wood, S.: Interactive home shopping: consumer, retailer, and manufacturer incentives to participate in electronic marketplaces. Journal of Marketing 61(3), 38–53 (1997)
18. Geyskens, I., Steenkamp, J.B.E.M., Kumar, N.: A Meta-Analysis of Satisfaction in Marketing Channel Relationships. Journal of Marketing Research 36, 223–238 (1999)
19. Hoque, A.Y., Lohse, G.: An Information Search Cost Perspective for Designing Interfaces for Electronic Commerce. Journal of Marketing Research 36, 387–394 (1999)
20. Childers, T.L., Carr, C.L., Peck, J., Carson, S.: Hedonic and Utilitarian Motivations for Online Retail Shopping Behaviour. Journal of Retailing 77, 511–535 (2001)
21. Kim, S., Stoel, L.: Apparel Retailers: Website Quality Dimensions and Satisfaction. Journal of Retailing and Consumer Services 11(2), 109–117 (2004)
22. Wilde, S.J., Kelly, S.J., Scott, D.: An Exploratory Investigation into E-Tail Image Attributes Important to Repeat, Internet Savvy Customers. Journal of Retailing and Consumer Services 11, 131–139 (2004)
23. Gorn, G.J., Chattopadhyay, A., Sengupta, J., Tripathi, S.: Waiting for the Web: How Screen Color Affects Time Perception. Journal of Marketing Research 51, 215–225 (2004)
24. Flavián, C., Guinalíu, M., Gurrea, R.: The Role Played by Perceived Usability, Satisfaction and Consumer Trust on Website Loyalty. Information and Management 43, 1–14 (2006)
25. Chen, H., Wigand, R.T., Nilan, M.S.: Optimal Experience of Web Activities. Computers in Human Behavior 15, 585–608 (1999)
26. Nielsen, J.: Usability Engineering. Morgan Kaufmann, San Francisco (1994)
27. Nielsen, J.: Usability 101 (2003),
    http://www.useit.com/alertbox/20030825.html
28. Muir, B.M., Moray, N.: Trust in Automation. Part II. Experimental Studies of Trust and Human Intervention in a Process Control Simulation. Ergonomics 39(3), 429–460 (1996)
29. Cristóbal, E.: El Merchandising en el Establecimiento Virtual: una Aproximación al Diseño y la Usabilidad. Esic Market 123, 139–163 (2006)
30. Palmer, J.W.: Web Site Usability, Design, and Performance Metrics. Information Systems Research 13(2), 141–167 (2002)
31. Baker, J.: The role of environment in marketing services: the consumer perspective. In: Cepeil, et al. (eds.) The Services Challenge: Integrating for Competitive Advantage, pp. 79–84. AMA, Chicago (1986)
32. Donovan, R.J., Rossiter, J.R., Marcoolyn, G., Nesdale, A.: Store Atmosphere and Purchasing Behaviour. Journal of Retailing 70(4), 283–294 (1994)
33. Spies, K., Hesse, F., Loesch, K.: Store Atmosphere, Mood and Purchase Behaviour. Journal of Research in Marketing 14, 1–17 (1997)
34. Turley, L.-W., Milliman, R.E.: Atmospheric Effects on Shopping Behavior: A Review of the Experimental Evidence. Journal of Business Research 49, 193–211 (2000)
35. Eroglu, S.A., Machleit, K.A., Davis, L.M.: Atmospheric Qualities of Online Retailing: a Conceptual Model and Implications. Journal of Business Research, Special Issue on Retail Strategy and Consumer Decision Research 54(2), 177–184 (2001)
36. Dailey, L.: Navigational Web Atmospherics. Explaining the Influence of Restrictive Navigation Cues. Journal of Business Research 57(7), 1–9 (2004)

37. Dailey, L.: Designing the World We Surf in: A Conceptual Model of Web Atmospherics. In: Brow, S., Sudharshan, D. (eds.) Proceedings of AMA Summer Educator's Conference, Chicago, Illinois (1999)
38. Nielsen, J., Molich, R.: Heuristic Evaluation of User Interfaces. In: Proceeding of the ACM Chicago 1990 Conference, Seattle, WA, April 1-5, 1990, pp. 249–256 (1990)
39. Nielsen, J.: Heuristic evaluation. In: Nielsen, J., Mack, R.L. (eds.) Usability Inspection Methods. John Wiley & Sons, New York (1994),
    http://www.useit.com/papers/heuristic/heuristic_list.html
40. Sutcliffe, A.: Heuristic Evaluation of Website Attractiveness and Usability. In: Interactive Systems: Design, Specification, and Verification, pp. 183–198. Springer, Heidelberg (2001)
41. Gehrke, D., Turban, E.: Determinants of Successful Website Design: Relative Importance and Recommendations for Effectiveness. In: Proceedings of the 32nd Hawaii International Conference on System Sciences, Hawaii, United States (1999)
42. Schubert, P., Selz, D.: Web Assessment – Measuring the Effectiveness of Electronic Commerce Sites Going beyond Traditional Marketing Paradigms. In: Proceedings of the 32nd Hawaii International Conference on System Sciences, Hawaii, United States (1999)
43. Gefen, D.: E-commerce: the Role of Familiarity and Trust. Omega, The International Journal of Management Science 28, 725–737 (2000)
44. Pereira, R.E.: Optimizing Human-Computer Interaction for the Electronic Commerce Environment. Journal of Electronic Commerce Research 1(1), 23–34 (2000)
45. Sutcliffe, A.: Assessing the Reliability of Heuristic Evaluation for Website Attractiveness and Usability. In: Proceedings of the 35th Hawaii International Conference on System Sciences, Hawaii, United States (2002)
46. Hwang, Y., Kim, D.J.: Customer Self-Service Systems: The Effects of Perceived Web Quality with Service Contents on Enjoyment, Anxiety, and E-Trust. Decision Support Systems 43(3), 746–760 (2007)
47. International Academy of Digital Arts and Sciences: The Webby Awards (2004), http://www.webbyawards.com/webbys/current.php?season=8# Commerce
48. World Best Enterprises: World Best Website Awards (2004), http://www.worldbest.com/shopping.htm
49. Hunt, B.: 10 Best-designed Web Sites in the World (2008), http://www.web-designfromscratch.com/10-best-designed-web-sites.cfm
50. Jarvenpaa, S.L., Tractinsky, N., Vitale, M.: Consumer Trust in an Internet Store. Information Technology and Management 1, 45–71 (2000)
51. Hong, W., Thong, J.Y.L., Tam, K.Y.: Designing Product Listing Pages on E-Commerce Websites: an Examination of Presentation Mode and Information Format. International Journal of Human-Computer Studies 61, 481–503 (2004)
52. Tractinsky, N., Lowengart, O.: Web-Store Aesthetics in E-Retailing: a Conceptual Framework and Some Theoretical Implications. Academy of Marketing Science Review 11(1), 1–18 (2007)
53. Fogg, F.J., Soohoo, C., Danielsen, D., Marable, L., Stanford, J., Tauber, E.R.: How do people evaluate a Web site's credibility?. Persuasive Technology Lab, Stanford University (2002)
54. Zhang, P., Von Dran, G.M., Small, R.V., Barcellos, S.: A Two Factor Theory for Website Design. In: Proceedings of the 33rd Hawaii International Conference on System Sciences, Hawaii, United States (2000)

55. Lee, W., Benbasat, I.: Designing an Electronic Commerce Interface: Attention and Product Memory as Elicited by Web Design. Electronic Commerce Research and Applications 2, 240–253 (2003)
56. Lorenzo, C., Mollá, A.: Website Design and E-Consumer: Effects and Responses. International Journal of Internet and Advertising 4(1), 114–141 (2007)
57. Zviran, M., Glezer, C., Avni, I.: User Satisfaction from Commercial Web Sites: The Effect of Design and Use. Information and Management 43, 157–178 (2006)
58. Richard, M.O.: Modelling the impact of Internet atmospherics on surfer behaviour. Journal of Business Research 58, 1632–1642 (2005)
59. European Interactive Advertising Association: EIAA Mediascope Europe 2007 Pan European Executive Summary (2007),
    http://www.eiaa.net/research/research-details.asp?SID=1&id=375&lang=6&origin=media-consumption.asp
60. Agarwal, R., Venkatesh, V.: Assessing a Firm's Web Presence: A Heuristic Evaluation Procedure for the Measurement of Usability. Information Systems Research 13(2), 168–186 (2002)
61. Huizingh, E.K.R.E.: The Content and Design of Web Sites: an Empirical Study. Information and Management 37, 123–134 (2000)
62. Keeney, R.: The Value of Internet to the Customer. Management Science 45(4), 533–542 (1999)

Service Architecture Design for E-Businesses: A Pattern-Based Approach

Veronica Gacitua-Decar and Claus Pahl

School of Computing, Dublin City University, Dublin 9, Ireland
{vgacitua,cpahl}@computing.dcu.ie

Abstract. E-business involves the implementation of business processes over the Web. At a technical level, this imposes an application integration problem. In a wider sense, the integration of software and business levels across organisations becomes a significant challenge. Service architectures are an increasingly adopted architectural approach for solving Enterprise Applications Integration (EAI). The adoption of this new architectural paradigm requires adaptation or creation of novel methodologies and techniques to solve the integration problem. In this paper we present the pattern-based techniques supporting a methodological framework to design service architectures for EAI. The techniques are used for services identification, for transformation from business models to service architectures and for architecture modifications.

1 Introduction

E-commerce, and more generally E-businesses, involve the implementation of business processes over the Web. The processes could span different organisations and include several software applications. At a technical level, the implementation of e-businesses imposes an enterprise applications integration (EAI) problem.

Nowadays, Service-oriented Architecture (SOA) is considered a promising architectural approach for EAI. Several methodologies such as [1],[2],[3] have been proposed for service architecture development. However, even though they provide useful guidelines, they are still maturing in aspects such as the providing of techniques promoting the automation of the architecture design process, and an adequate meta-modelling framework for capturing both, business and software aspects.

In [4], we developed a framework for designing service architectures for EAI. The framework is driven by business models and exploits patterns to support the design of architectures. Software patterns are considered in the software community as architectural abstractions representing encapsulated practical design knowledge [5]. Similarly, business reference models and business patterns provide encapsulated modelling knowledge [6]. The methodological framework guides the creation of architectures while incorporating the business and software dimensions of the integration problem. The approach provides explicit traceability between business and software modelling elements and uses patterns as central elements to provide improved changeability characteristics to the resultant architectures.

The objective of this paper is to describe the pattern-based techniques required for such a development framework. The framework is structured in a layered architecture,

G. Psaila and R. Wagner (Eds.): EC-Web 2008, LNCS 5183, pp. 41–50, 2008.
© Springer-Verlag Berlin Heidelberg 2008

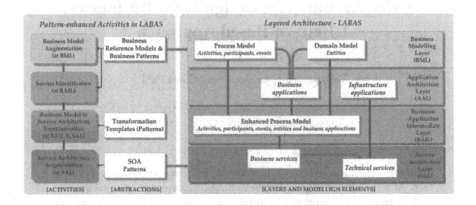

Fig. 1. LABAS and pattern-enhanced activities of the design process

their main elements are summarised in section 2. The pattern-based techniques are described in section 3. We finalise the paper with a discussion, revision of related work and conclusions in sections 4 and 5.

2   Layered Architecture: Layers and Architecture Abstractions

The framework provides a layered architecture called LABAS[1] for structuring the EAI problem [4]. The development of architectures solutions with LABAS involves the incremental transformation from business models into service architectures. The transformation is supported by architectural abstractions and pattern-based techniques. Fig.1-right side depicts the architecture layers and its elements. On the left side, the involved architecture abstractions and the main steps of the development processes are shown.

Layers. Layers in LABAS are the Business Modelling Layer (BML), the Application Architecture Layer (AAL), the Business-Applications Intermediate Layer (BAIL) and the Service Architecture Layer (SAL).

- *BML* is the container for the business model that represents the business context of the integration problem. Models in BML are expressed in an enhanced BPMN notation that includes domain model elements.
- *AAL* is the container for the application components supporting the business processes in BML. AAL is structured in an applications architecture model.
- *BAIL* is the container for an enhanced business process model that integrates the elements from BML and AAL, supported an explicit traceability model.
- *SAL* is the container for the service architecture that solves the integration problem. Services are categorised in two main types: *business services* and *technical services.*

---

[1] Layered Architecture for Business, Applications and Services.

Architectural Abstractions. In LABAS, business reference models, business patterns and SOA patterns are the main considered architectural abstractions.

Business reference models provide a standard model and decomposition of a business domain. They are originated from experience and combined with business patterns, they can constitute business reference architectures.

Business patterns are micro-models describing standard decompositions of business reference models. Similarly to software reference architectures, the business reference architectures constrain the composition of business patterns. Two main business patterns kinds are considered: *process patterns* and *domain patterns*. A further categorization regarding specific domains might be also considered in LABAS, but it is not mentioned in the paper.

SOA patterns are software design patterns that provide solutions for technical aspects of service architectures, such as service invoking, service composition, security, among others.

Business and SOA patterns have a semantic dimension and a structural dimension, however process patterns add a behavioural dimension. In this paper, we will refer only to structural aspects of patterns.

Layered architecture implementation. A meta-model and profile for the layers in LABAS is provided. Most LABAS constructs are mapped to UML 2.0 constructs. Traces in BAIL are an integral part of the LABAS profile and follows the trace-tagged traceability meta-model from [7]. The LABAS meta-model and profile for SAL are based on a proposal for the UML Profile and Meta-model for Services (UPMS) provided by the OMG.

Patterns are implemented and organised in pattern catalogues. Each pattern contains information organised in a *pattern template*, which includes sections such as intent, motivation, participants, consequences, associated quality attributes, among others. Usually, pattern templates involves textual descriptions [8]. We add the models representing the patterns into the pattern templates. The models representing the patterns use elements from the LABAS profile. This approach allows the interchange of the pattern catalogue and its elements, encouraging in this manner, the using of patterns as tool-supported modelling constructs.

Note that the implementation of the pattern-based techniques -described in the next section- will be part of a plug-in for a standard UML modelling tool. The plug-in is complemented with the LABAS profile, compliant with the LABAS meta-model.

3   Pattern-Based Techniques to SOA Design

The pattern-based techniques in LABAS provide support to business analysts and software architects to incrementally design service architectures. The main activities supported by pattern-based techniques are depicted in Fig.1-right side, and encompass:

– *Business model augmentation* involves the addition of business patterns into the business model. Recommendation of the using of patterns, pattern instantiation and patterns combination are techniques supporting this activity.

- *Business and technical service identification* involves the analysis of business models and their relations with the supporting software applications to define the services of the architecture solution. Patterns identification and patterns matching are two techniques aiding this activity.
- *Business model to service architecture transformation* incrementally generates the service architecture solution. This activity is supported by pattern-based transformation templates.
- *Service architecture augmentation* incorporates SOA patterns to provide solutions to technical issues such as communication between services, security, services distribution, among others. This activity is analogous to business model augmentation.

3.1   Business Model and Architecture Augmentation

Patterns represent experience-based solutions that can be applied to improve the quality of designs. The LABAS framework provides a repository of patterns in the form of pattern catalogues and the techniques that allows their use. This section describes the techniques allowing the use of patterns to augment business models and architectures.

Pattern recommendation. Less experienced designers might not be aware that a pattern might be applied to improve the quality of their designs. The utilisation of patterns into designs requires the recognition of the associated design problem. Tool support for pattern recommendation requires to increase the degree of automation for the pattern-problem identification[2]. If the pattern problem is expressed in terms of elements and relations of the design model, then a design problem could be systematically searched, and once located, it can be suggested to the designer to allow the subsequent instantiation of the associated pattern-solution. The modelling of the pattern problem is a key issue for this task [9].

In LABAS, business models, architectures and patterns are represented as graphs to support the business model and architecture augmentation activity. From a structural point of view, the identification of the pattern-problem is supported by identifying the structural features that individually characterise each pattern-problem model. These structural features are represented as subgraphs of the graph representing the pattern-problem model. The systematic searching of the structural features is supported by graph matching techniques.

Pattern comparison. A design problem could have more than one pattern solution associated. In this case, two or more patterns require comparison. Comparison is supported in LABAS with information about quality attributes associated to patterns. This information is encapsulated in the pattern consequences section of the pattern template.

Pattern modification. The documentation of a pattern-solution provides a *generic* solution. In order to instantiate a pattern in a specific model, the generic pattern-solution might require modifications to become suitable for the specific model. The preservation of the pattern properties requires that only allowed modifications can be done. The

---

[2] Note that a pattern description in a pattern template includes a generic problem-solution pair and its variations.

allowed modifications are implemented in a set of techniques that can be applied to modify patterns. For instance, a basic technique to modify the *mediator* pattern [8] involves the increasing or decreasing of *colleague* elements. After modification of the pattern, validation techniques are applied.

Pattern instantiation. Pattern instantiation involves the creation of elements and relations from the pattern model into a design model, and/or the merging of pattern elements with design model elements. In LABAS, patterns can be instantiated to augment models at BML and SAL layers.

In order to increase the degree of automation, once selected a pattern, its instantiation can be carried out through the implementation of a graph transformation rule $p : L \longrightarrow R$ [10], where L corresponds to the graph that represents the pattern-problem, and R corresponds to the graph that represents the pattern-solution. The transformation rule p allows the transformation from a graph that represents the model or architecture with a design problem, into a graph that represents the model or architecture with the instantiated pattern-solution.

Pattern combination. Architecture designs and business models can contain more than one pattern instance. Patterns can be combined to provide a design solution with a larger scope. Let us say that PAT_i and PAT_j are two patterns in a pattern catalogue, and $G_i = (V_i, E_i)$, $G_j = (V_j, E_j)$ are graphs[3] representing those patterns. Different types of pattern combinations might occur, depending on if $V_i \cap V_j = \emptyset$, where $\emptyset$ is an empty set; or if $V_i \cap V_j \neq \emptyset$. If the latter case occur and $PAT_i \subset PAT_j$, then PAT_i is *embedded* within PAT_j. In other cases the *union* of patterns is utilised. Union and embedding are to basic operations for pattern combination [11].

Requirements to combine patterns could exceed the capabilities of basic techniques. Only as an illustration, we use an analogy with relational algebra. Basic operations as projection (π) and selection (σ) in relational algebra are not enough in some practical uses for data base queries. Composition of operators is a solution to the restrictions of the basic operations, e.g. $\pi_A(\sigma_P(r))$ [4] allows the projection on a selected set of tuplas of a relation r. Analogously, the combination of pattern techniques provides a medium to satisfy more complex requirements.

Combination of patterns could interfere with the expected contributions that each pattern provides separately. An important issue in pattern combination is to verify that individual pattern consequences are preserved. This is difficult to ensure before implementation. However, at design-time, it is possible to analyse the potential interferences between associated quality attributes of each pattern.

3.2 Pattern-Based Service Identification and Incorporation

The explanation of this section uses an example extracted from a case study in [4]. The case study involves a billing and payment process, representing a typical process where customers and businesses interact. The example is focused on a payment transaction

---

[3] V_i and V_j are the set of graph vertices. E_i and E_j are the set of graph edges in G_i and G_j.

[4] A: attributes where the relation r is projected. P: logic predicate satisfied by tuplas in r.

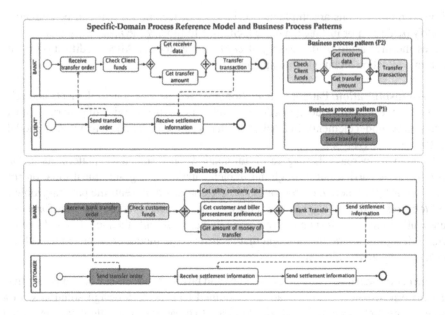

Fig. 2. Bank's transfer process modelled with the BPMN notation - right side. Business process reference model and patterns - left side.

- bank's transfer. Customers can send a bank's transfer order to pay their bills. After-wards, the bank verifies the funds in the payer's account and executes the transaction. After the transaction is completed, the bank sends the settlement information to the payer. The customer in turn sends the settlement information to the biller. Fig.2-bottom shows the high level business process. A simplified reference model and two simplified examples of business process patterns (P1 and P2) are shown at the top of the figure.

Business Service Identification. The identification of business patterns facilitates the identification of reusable business services. Business patterns are reusable sections of business models, and they are a common denominator among different organisations and within the same organisation that is changing over time. Changeability is related to the ease of an architecture to change, but also with the ability of the architecture to remain invariant after a change agent acts [12]. The definition of business services in LABAS takes into account the latter characteristic.

In order to foment the automation of business pattern identification, *pattern matching* techniques are utilised. From a structural point of view, the pattern matching technique is based on the matching of a graph (G_{PAT}) representing the pattern over a graph G_{BM} representing the business model. The granularity of the pattern is such that $G_{PAT} \subseteq G_{BM}$. To identify a business pattern, and consequently a business service, an algorithm searches for the sub-graph G_{PAT} within the graph G_{BM}. Fig.2 shows a simplified schema of a reference process model, patterns and a business model containing those patterns.

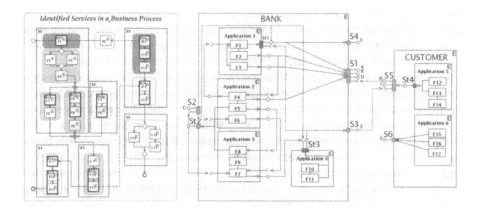

Fig. 3. Business and technical services identified in a enhanced business process model - left side. The related service architecture - right side.

Technical Service Identification. The identification of common flow structures in process models from BAIL is the basis for the identification of technical services. In order to identify common flow structures the process models from BAIL are decomposed until the activities with a one to one relation with applications components from AAL are reached. The flow structures have explicit traces with application components from AAL. The traces represent invocations to functionality of applications from AAL or responses from applications to fulfill information requirements of steps in the process flow. The identification of technical services across process models pursues the fundamental concept of reuse in SOA. The identification of common control flow structures is supported in LABAS by graph partitioning techniques. Graph representing enhanced process models have information about the types of their elements, therefore control flow structures involving certain types of elements can be further categorised as different technical services types, for instance: data aggregation, calculations, among others.

The Fig.3 sketches a simplified decomposition of the process model of the Fig.2. Note that the proper modelling notation of LABAS is not used because of space constraints. S1 to S6 represent the identified business services. S1 and S5 were defined through business pattern matching. St1 to St4 are the defined technical services through the identification of common flow structures in BAIL. Technical services can be reused by business services. Note that in order to simplify the illustration of technical services, only simple flow structures were depicted in the Fig.3. More complex structures involving other control flow structures, for instance decisions, joins, among others, can also be considered.

3.3 Business Model to Service Architecture Transformation

The transformation from business models into service architectures involves the tranformation of identified software services in BAIL to service elements in SAL. The transformation uses a pattern-based *transformation template* that maps, at meta-modelling level, architectural abstractions from BAIL into SAL. The relations among BAIL elements are preserved after the transformation. These relations provide information about the flow dependencies between business services. Fig.3-right side shows the software architecture model generated from the BAIL model depicted in the left side.

Note that several model driven development approaches have followed a strategy of direct translation from business modelling constructs to software constructs, e.g. direct transformation from BPMN-to-BPEL constructs. However, business models could contain sections that cause deadlocks and other problems for the process execution [13]. As mentioned previously, in LABAS, the transformation from business models to service architectures is based on pattern-based transformation templates. An advantage of using transformation templates is that they can be designed to provide only error-free transformations. Nevertheless, this require a previous step to refine the business process model to match with the business model section of the transformation template. In [14], a control-flow pattern approach for BPMN-to-BPEL translation is presented. The transformation templates in LABAS follows a similar approach, but beyond control-flow structures.

4   Related Work and Discussion

Different pattern-based techniques have been proposed for analysis, design and evolution of architectures. In [15], a systematic method to select patterns using language grammars and design space analysis is introduced. In a similar approach, a method to design solutions for transactional workflows for SOA is presented [16]. The design method models alternative solutions to recurrent architectural decisions, as patterns and primitives. These architecture abstractions are, in turn, mapped to technology-based solutions. Both approaches follow a manual-based approach for pattern selection and instantiation. In our case, the aim is provide support for (semi-)automation.

Sets of patterns are normally part of organised collections named pattern languages. Pattern languages allow regulated combinations that extends the reach of individual patterns [17]. In [18] a pattern language for process-oriented integration of software services is presented. We focus not only on patterns at software level, but also at business level and we investigate the relation of patterns at those two levels.

After software implementation, changes on it might interfere with the previously applied design patterns. In [19] a graph-transformation approach to pattern level design validation and evolution is presented. Only structural aspects are reviewed. In [20], current limitations of patterns on evaluating their impact on quality attributes is presented, however the interaction of the pattern's consequences have not been discussed. In [21], an ontological-based approach for modelling architecture styles is presented. Style modifications and combinations among them are introduced. Relations between quality requirements and modelling of styles are investigated. We are currently extending the ideas in [21].

A key step in the SOA development processes is to define what are the involved software services [1]. We presented an approach to service identification based on business models and related reference models. Development and maintenance of business and software architecture models, together with associated reference architectures and reference models, have been encouraged with the increasing use of enterprise architecture frameworks, such as for example the Zachman framework [22].

In this paper we have neglected process simulation and semantic considerations for pattern matching and pattern identification techniques. However, the integration of behavioral aspects and semantic aspects will be investigated. The integration of structural, behavioral and semantic aspects could consider the directions followed in [23] and [24]. To evaluate the architecture solutions created with LABAS we have considered the use of the Architecture-Level Modifiability Analysis (ALMA) method [25]. In [4] we have demonstrated the use the proposed layered architecture and discussed the use of ALMA for a modifiability analysis.

5   Conclusion

Traditionally, the creation of architectures have only focused on structural descriptions. Instead, we focus on processes and constrained architectural descriptions. The continual rise of abstraction in software engineering approaches was a central driver, placing the notion of patterns at business domain level and focusing on its subsequent transformation to a service architecture.

In this paper we have presented the notational elements and pattern-based techniques used in our methodological framework (LABAS) to design service architectures for EAI. The presented pattern-based techniques are utilised for software service identification, for business model to service architecture transformations and for architecture modifications. The LABAS approach has as one of its ultimate goals, the creation of service architecture solutions with improved changeability characteristics, while maintaining coherence between the business model and the software architecture. The explicit traceability between elements of the different layers of LABAS contributes to the coherence between the business level and the software level. Changeability characteristics of the architecture solutions are improved by using patterns and the pattern-based techniques described in this paper.

From the point of view of designing architectures, the integration of business aspects and software aspects required to implement E-businesses involves three main dimensions: structure, behaviour and semantic. The consideration of these three dimensions in a single framework to design architectures is challenging since these three dimensions have been mostly investigated separately and with different formalisms. We aim to integrate these dimensions to provide an integral support to the designers interested in the developing of service architectures.

References

1. Erl, T.: Service-oriented architecture: Concepts, Technology, and Design. Prentice Hall, Englewood Cliffs (2004)
2. Papazoglou, M.P., van den Heuvel, W.J.: Service-oriented design and development methodology. Int. J. of Web Engineering and Technology (IJWET) 2, 412–442 (2006)
3. Arsanjani, A.: Service-oriented modeling and architecture (2004)
4. Gacitua-Decar, V., Pahl, C.: Business model driven service architecture design for enterprise application integration. In: ICBIIT 2008 (2008)

5. Bass, L., Clements, P., Kazman, R.: Software Architecture in Practice, 2nd edn. Addison-Wesley Professional, Reading (2004)
6. Fettke, P., Loos, P.: Reference Modeling for Business Systems Analysis. IGI (2006)
7. Baelen, V.v., Berbers, J.: Traceability as input for model transformations. In: ECMDA Traceability Workshop (ECMDA-TW), Haifa, Israel (2007)
8. Gamma, E., Helm, R., Johnson, R., Vlissides, J.: Design Patterns: Elements of Reusable Object-Oriented Software. Addison-Wesley Professional Computing Series (1995)
9. Kim, D.K., Khawand, C.E.: An approach to precisely specifying the problem domain of design patterns. J. of Visual Languages and Computing 18(6), 560–591 (2007)
10. Corradini, A., Montanari, U., Rossi, F., Ehrig, H., Heckel, R., Lowe, M.: Algebraic approaches to graph transformation. Handbook of Graph Grammars and Computing by Graph Transformation 1, 163–245 (1997)
11. Gomes, M.C., Rana, O.F., Cunha, J.C.: Pattern operators for grid environments. Sci. Program. 11(3), 237–261 (2003)
12. Ross, A., Rhodes, D., Hastings, D.: Defining changeability: Reconciling flexibility, adaptability, scalability, modifiability, and robustness for maintaining system lifecycle value. Journal of Systems Engineering (accepted, 2008)
13. Koehler, J., Gschwind, T., Kuster, J., Pautasso, C., Ryndina, K., Vanhatalo, J., Volzer, H.: Combining quality assurance and model transformations in business-driven development. In: AGTIVE 2007. LNCS, vol. 5088. Springer, Heidelberg (2008)
14. Ouyang, C., Dumas, M., ter Hofstede, A.H.M., van der Aalst, W.M.P.: Pattern-based translation of bpmn process models to bpel web services. International Journal of Web Services Research (2007)
15. Zdun, U.: Systematic pattern selection using pattern language grammars and design space analysis. Software Practice and Experience 37(9), 983–1016 (2007)
16. Zimmermann, O., Grundler, J., Tai, S., Leymann, F.: Architectural decisions and patterns for transactional workflows in soa. In: Krämer, B.J., Lin, K.-J., Narasimhan, P. (eds.) ICSOC 2007. LNCS, vol. 4749, pp. 81–93. Springer, Heidelberg (2007)
17. Buschmann, F., Henney, K., Schmidt, D.C.: Pattern-Oriented Software Architecture: On Patterns and Pattern Languages. Wiley and Sons, Chichester (2007)
18. Hentrich, C., Zdun, U.: Patterns for process-oriented integration in service-oriented architectures. In: EuroPLoP 2006, Irsee, Germany, pp. 1–45 (2006)
19. Zhao, C., Kong, J., Dong, J., Zhang, K.: Pattern-based design evolution using graph transformation. J. of Visual Languages and Computing 18(4), 378–398 (2007)
20. Harrison, N.B., Avgeriou, P.: Leveraging Architecture Patterns to Satisfy Quality Attributes. In: Oquendo, F. (ed.) ECSA 2007. LNCS, vol. 4758, pp. 263–270. Springer, Heidelberg (2007)
21. Pahl, C., Giesecke, S., Hasselbring, W.: An ontology-based approach for modelling architectural styles. In: Oquendo, F. (ed.) ECSA 2007. LNCS, vol. 4758, pp. 60–75. Springer, Heidelberg (2007)
22. Sowa, J.F., Zachman, J.A.: Extending and formalizing the framework for information systems architecture. IBM Syst. J. 31(3), 590–616 (1992)
23. Ehrig, M., Koschmider, A., Oberweis, A.: Measuring similarity between semantic business process models. In: APCCM 2007, Australia, vol. 67, pp. 71–80 (2007)
24. Martens, A.: Simulation and equivalence between bpel process models. In: Proc. of the Design, Analysis, and Simulation of Distributed Systems Symposium (DASD 2005) (2005)
25. Bengtsson, P., Lassing, N., Bosch, J., van Vliet, H.: Architecture-level modifiability analysis (alma). Journal of Systems and Software 69(1-2), 129–147 (2004)

An Event-Based Model for the Management of Choreographed Services*

Liliana Ardissono, Roberto Furnari, Anna Goy, Giovanna Petrone,
and Marino Segnan

Dipartimento di Informatica - Università di Torino
Corso Svizzera 185, 10149 Torino - Italy
{liliana,furnari,goy,giovanna,marino}@di.unito.it
http://www.di.unito.it

Abstract. We propose a mediation framework supporting a flexible management of choreographed services. The framework separates the management of the business logic of a service from the message flow details reported in the choreography specification. This abstraction step is achieved by introducing an event-driven management of Web Services and by exploiting a rich coordination context which includes business data and synchronization information. By separating activity management issues from communication details, our framework facilitates the management of interaction protocol mismatches between Web Services.

1 Introduction

In Web Service composition, choreographies are introduced to specify applications whose business logic is managed in a decentralized way; see [1]. A choreographed service results from the cooperation of multiple Web Services (WSs) which coordinate the activities to be performed via message passing.

A choreography specification represents a contract to be accepted by the WSs which want to cooperate to the management of a composite service. The contract enables the WSs to agree on the activities to be performed, preventing run time errors and violations of the business logic of the composite service. However, in order to participate as a role filler in the service, a WS has to adapt its own interaction protocol to the choreography specifications, by solving the occurring mismatches. For instance, there might be differences in operation names, parameter lists and message flow. This means that a WS participating to more than one composite service might be provided with different WSDL [2] interfaces. Moreover, for each choreography, an adapter should be developed in order to enforce the specified order in the generation and reception of messages.

In order to address such issues, we propose an event-driven choreography management framework supporting the mediation of interaction protocols within a composite service. Our model is based on the idea that, during the execution of

---

* This work was partially funded by projects WS-Diamond (IST-516933) and QuaD-RAnTIS (MUR).

G. Psaila and R. Wagner (Eds.): EC-Web 2008, LNCS 5183, pp. 51–60, 2008.

a composite service, it is possible to abstract from several message flow details specified in the choreography, provided that the data dependencies and synchronization constraints imposed by the business logic of the service are respected. In order to support the cooperation between Web Services and their synchronization, our model introduces a Choreography Coordination WS which manages the Choreography Coordination Context by collecting business data and synchronization information and by propagating them to the interested Web Services according to the Publish and Subscribe pattern. Our framework also enforces an event-driven execution of activities [3] within WSs, in order to enable them to autonomously operate, as soon as they are provided with the data items they need, and the related synchronization constraints are satisfied.

In the following, Section 2 presents the motivations of our work and Section 3 describes our choreography management model. Section 4 describes the event-based representation of a choreography that is the basis of our proposal. At last, Section 5 positions our work in the related work and concludes the paper.

2   Motivations of Our Work

A choreography specification describes the business logic of a composite service in a message-oriented way, by specifying the expected interaction flow between the cooperating WSs. As messages invoke WSDL operations, this specification enforces the execution of activities, therefore supporting the WS coordination. However, the WSs participating to an application might offer operations which differ from those specified in the messages of the reference choreography. Thus, in order to fill a role within a composite application, the interaction protocol of a WS has to be adapted to the message flow specified in the choreography.

We assume that the choreography and the WS interaction protocols are described as UML activity diagrams. This notation is suitable for describing the interaction flow. In fact, UML activity diagrams can be mapped to Petri Nets, which are a reference model for the specification of processes; see [4].

Even though a Web Service complies with the representation and meaning of the business data handled in the application, various protocol mismatches are possible. For instance, the names and parameter list of the WSDL operations might differ from those occurring in the choreography specification. Moreover, the expected number and order of messages might differ from the interaction protocol of the WS. Although ad-hoc adapters can be developed to solve protocol mismatches, a general solution to the individual Web Services interaction protocol adaptation should be identified.

Indeed, the information to be specified in order to support the WS coordination within a choreographed service consists of: the business data to be handled, the operations to be performed on data items, the data dependencies between operations and possibly other pre-conditions on operations. However, a choreography specification embeds this information in a detailed message flow specification which conveys additional constraints, such as the signatures of the operations. We claim that a general run-time model for choreography management can be developed

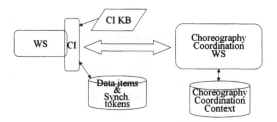

Fig. 1. Architecture of our Choreography Coordination Model

by abstracting from the flow details imposed by message-oriented coordination and by focusing on data-dependencies and synchronization. This approach deeply changes the role of the choreography specification, which does not represent any more the run-time behavior of the cooperating WSs. In fact, the choreography is used when the composite service is set up, as a reference to build the run-time model of the service and to help solving mismatches between the protocols of the individual WSs filling a particular role within the choreography.

3   Choreography Management Model

3.1   Architecture

We propose to manage the cooperation among WSs by handling a Choreography Coordination Context which supports the coordination of the WS activities within the composite application. Figure 1 depicts the architecture of our choreography management model:

- The Choreography Coordination WS (ChCoordWS) manages the Choreography Coordination Context of the application, which maintains:
    • Basic coordination information; e.g., the identifier of the choreography instance which the cooperating Web Services participate to; see [5].
    • The data items to be shared between WSs (business data).
    • The synchronization constraints that have to be satisfied during the execution of the application. As described in Section 4, the choreography specification and the interaction protocol of the WS are translated to an event-based representation where synchronization constraints are represented as tokens. Thus, the satisfaction of a synchronization constraint is described by making the corresponding token available in the Choreography Coordination Context.
- In order to enable the cooperating WSs to perform activities autonomously, our model replaces the message-based invocation of WSDL operations with an event-driven action execution [3]. Each cooperating WS is thus wrapped by an adapter, the Communication Interface (CI), which activates the execution of the services within the WS, on the basis of the available context

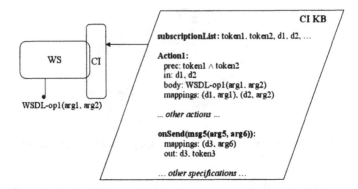

Fig. 2. Architecture of a WS supplier

information and of the existing synchronization constraints. The CI blocks the interaction with the other WSs in order to avoid the direct invocation of WSDL operations; moreover, it manages the interaction between the WS and the ChCoordWS.

The ChCoordWS collects business data and synchronization information and propagates them to the CIs of the interested Web Services according to the Publish and Subscribe pattern, supporting distributed caching. Each CI receives from the ChCoordWS the information it had subscribed for, as soon as it becomes available. Moreover, when the wrapped WS generates some data items, or it satisfies a synchronization constraint, the CI publishes such information to the ChCoordWS, which notifies the interested WSs.

Notice that we have opted for a centralized management of the Context information because it supports the data consistency management (e.g., in the propagation of changes in the business data). In contrast, we have adopted a distributed management of the Web Service activities to minimize the amount of application dependent information which the ChCoordWS has to manage.

We are implementing our model by exploiting Web Services and message handlers for the development of the CIs. Moreover, we are implementing the ChCoordWS by exploiting the GigaSpaces [6] middleware, which supports the propagation of data items according to the Linda [7] tuple space model. GigaSpaces takes care of logging the publication and notification messages occurring during the execution of the application; as such, it provides a trace which can be utilized for error management purposes.

3.2   Action-Based WS Representation

The event-driven activity execution is based on a representation of activities (WSDL operations) as the actions of an autonomous agent. The CI has a knowledge base (CI KB) storing the information needed to support the WS execution; see Figure 2. The CI KB stores the following types of information:

– The specification of the actions which the WS has to perform during its execution as the filler of a role of the application. In a composite application, a WSDL operation might be invoked more than once. Thus, given a WSDL operation, for each possible invocation, an $\mathtt{Action}_x$ slot is defined to describe the invocation setting. The slot specifies the following information:

1. **in**: this field specifies the business data items (as described in the choreography specification) needed as input parameters for the execution of the WSDL operation.
2. **out**: this field specifies the business data items and synchronization tokens produced by the operation (if any).
3. **mappings**: this field maps the business data to the input/output arguments of the operation.
4. **prec**: this field describes the precondition of the action, i.e., the synchronization constraints which must be satisfied in order to enable the execution of the operation. The preconditions are represented as boolean expressions on synchronization tokens.

For instance, **Action1** in the figure describes preconditions and input parameters of the WSDL operation named **WSDL-op1**. The operation can be executed only after the synchronization constraints represented by *token1* and *token2* have been satisfied. Moreover, the arguments of the action (*arg1* and *arg2*) have to be bound to the *d1* and *d2* business data items.

– The CI KB also specifies the subscription list of the WS (**subscriptionList**). This is the list of business data and synchronization tokens relevant for the WS execution (i.e., occurring in the input parameters of actions and in their preconditions).

– Finally, for each outbound message *msg* belonging to the interaction protocol of the WS, an **onSend**(*msg*) slot specifies in its **out** field the business data items and the synchronization tokens to be published in the Choreography Coordination Context when the message is generated (plus the mappings between business data and message parameters). For example, the CI KB in Figure 2 specifies that, when the WS generates **msg5**(*arg5, arg6*), the *d3* data item and the *token3* synchronization token must be published.

3.3   Event-Driven WS Execution

The life cycle of a choreographed application includes an initialization and a management phases. At initialization time, the CI of each cooperating WS registers in the Choreography Coordination Context by sending a registration request to the ChCoordWS, according to the WS-Coordination specifications [5].

At runtime, the CI mediates the interaction between the WS, the ChCoordWS and the other cooperating WSs by receiving the messages from the ChCoordWS and by intercepting the outbound messages of the WS.

– When a WS, following its own interaction protocol, tries to invoke a WSDL operation on another WS, the CI absorbs the outbound message. Then, the CI extracts from the message the values of the output parameters and it generates the synchronization tokens, according to the **onSend** specifications in

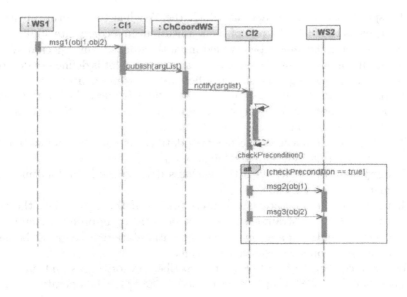

Fig. 3. Messages exchanged by WSs at choreography management time

the CI KB. Finally, it publishes the values in the Choreography Coordination Context, by notifying the ChCoordWS. We have depicted this interaction in Figure 3:[1] WS1 tries to invoke operation *msg*1 on the *obj*1 and *obj*2 actual parameters (`msg1(`*obj1*`, `*obj2*`)`); CI1 captures the message and publishes *obj*1, *obj*2 and the related synchronization tokens, by sending ChCoordWS the `publish(`*arglist*`)` message.

- When the ChCoordWS receives some data items or tokens, it stores them in the Choreography Coordination Context. Then, it propagates them to the CIs of the subscribed Web Services by sending a multicast `notify(`*arglist*`)` message.

- When a CI receives a set of data items and/or synchronization tokens, it stores them locally and checks whether any `Action` of the Web Service can be performed. If any `Action` has all the input parameters instantiated, and its preconditions are satisfied, the CI invokes the associated WSDL operation on the WS and retrieves the result. In Figure 3, we have depicted this situation by showing that, when receiving *obj*1 and *obj*2, CI2 invokes, respectively, the `msg2(`*obj*1`)` and `msg3(`*obj*2`)` operations on WS2.

By mediating the interaction between WSs, our model solves several protocol mismatch problems; e.g.:

- Mismatches in the operation names, number of parameters, and number of operations to be invoked; e.g., a WS might offer a WSDL operation performing

---

[1] In this example, for simplicity, we assume that business data items and local parameters of the interaction protocols have the same names.

a complex operation, but the client might try to separately invoke more than one operation for that purpose (or viceversa).

- Unexpected or missing acknowledgment messages; e.g., a WS might wait for an acknowledgment before continuing the execution of activities, but the interaction protocol of the client might not include such a message; viceversa, the client might send a message which should be ignored by the WS provider.
- Mismatches concerning the expected senders of messages; e.g. a WS might expect to be invoked by a certain client in order to perform an operation, but in the composite application another WS might send the invocation message. In our model, the propagation of business data hides the identity of the WS which has produced it; therefore, the mismatches involving the source or destination of messages can be ignored, provided that the needed business data is produced by other WSs.

4   Setting Up a Choreographed Application

In order to participate to a composite application, a WS must be wrapped by a CI and the CI KB must be configured according to the interaction protocol of the WS and the choreography specifications. By mapping the interaction protocol to the choreography messages, a correspondence is defined between the WS execution and the progress in the execution of the application. The mapping includes the following steps:

- First, the parameters of the messages occurring in the interaction protocol of the WS have to be mapped to the business data specified in the choreography. This task produces the *data mapping table*.
- Second, the inbound and outbound messages of the interaction protocol have to be mapped to the corresponding choreography messages (or sets of messages). If there is no correspondence, the messages must be positioned in a point of the choreography where they can be handled. This task produces a *message mapping table* which relates (possibly empty sets of) messages of the WS to (possibly empty sets of) choreography messages.

Given the mapping tables, the CI KB can be generated by defining an `Action` slot for each inbound message and an `onSend` slot for each outbound message of the WS interaction protocol. In order to facilitate the generation of such slots, we assume that both the choreography specification and the interaction protocol of the WS are translated to an internal format which makes synchronization information explicit: the *token-based representation*.

4.1   Token-Based Choreography/Protocol Representation

The UML activity diagrams describing a choreography specification and an interaction protocol can be translated to their token-based representations by adding an output synchronization token for each activity node of the diagrams, i.e., for each message.[2] The association of a token to a message means that, when the message is sent/received, the related synchronization constraint is satisfied.

However, this algorithm introduces redundant tokens because it does not take into account data dependencies between messages. Two types of dependencies can be considered in order to simplify the token-based representation: a) the dependencies among the messages producing data items and those having such parameters as input ones; b) the dependencies between messages which change the value of the data items. If this kind of information is specified (e.g., in WS-CDL [8]), the token-based representation can be simplified automatically; otherwise, the redundant tokens must be manually removed.

4.2   Generation of the CI KB

Starting from the mapping tables and the token-based representations of the interaction protocol and of the choreography specification, the CI KB can be automatically generated. Specifically:

- For each expected invocation of a WSDL operation, an Action slot has to be defined:
  - The in and out fields are specified by retrieving input and output parameters from the WSDL operation and by replacing them with the corresponding business data, as reported in the *data mapping table*. The same information is utilized to specify the mappings field.
  - The preconditions have to include the synchronization tokens which must be available when the WSDL operation can be performed, according to the local interaction protocol and to the choreography specification. The idea is that of including in the preconditions the synchronization information imposed by the choreography and to ignore local tokens, unless these represent additional synchronization constraints to be satisfied. Thus, the preconditions must include the input tokens of the (set of) choreography messages to which the Action corresponds. Moreover, in some cases (e.g., where there is not a one-to-one mapping between WSDL operations and choreography messages), the preconditions must also include some local tokens aimed at synchronizing the operations within the interaction protocol of the WS.[3]
- For each outbound message *msg* of the WS, an onSend(*msg*) slot has to be defined in the CI KB. The slot must specify the data items generated or modified by the WS; i.e., the output parameters of the message. Moreover,

---

[2] Moreover, as far as the choreography is concerned, two output tokens must be added for each decision node where the actor making the decision differs from the actors sending the first messages after the decision.

[3] The presence of additional synchronization constraints can be associated to unsolvable protocol mismatches; e.g., mismatches causing deadlocks in the choreography. We assume that the WS administrator takes care of checking that the business logic of the WS does not violate any constraints of the choreographed application; therefore, the additional constraints that we consider here are not problematic.

the slot must include the synchronization tokens to be generated by the CI, if any. The specification of the fields of the slot is similar to the one described for the `Action` slots.

- The subscription list is composed of the input parameters of the `Actions` and of the synchronization tokens appearing in their preconditions.

5   Conclusions and Related Work

We have presented a choreography coordination model which addresses the adaptation of Web Services' (WSs) to the business logic of the application, and the management of interaction protocol mismatches. The model is based on an event-driven execution of WS suppliers and on the introduction of a Choreography Coordination WS managing a shared coordination context.

Being event-driven, our model bears some relation to a Complex Event Processing architecture, but we focus on the execution of a specific choreography where applications with complex protocols interact with each other, not on the activation of applications in response to the detection of complex events.

The model presented in this paper extends conversational models, initially defined to support one-to-one communication between Web Services (e.g., [9,10,11]), to the management of many-to-many conversations.

Although several choreography specification languages have been defined (e.g., WSCI [10] and WS-CDL [8]), to our knowledge, no choreography management model has been introduced which supports a flexible and lightweight execution of a choreographed application. Our proposal also tries to provide an answer to this need.

Some mediation frameworks have been designed in order to enable the interaction between WSs; e.g., WSMO [12], WSMX [13], or the proposal by Benatallah and colleagues [14]. However, such frameworks are mainly focused on data mediation and they propose complex solutions as far as interaction protocol mediation is concerned. For instance, WSMX [13] introduces an external choreography management service which, similar to the orchestrator of a composite application, monitors the evolution of the execution state of the application and instructs participants about how to continue the interaction.

In other projects, the overhead of centralized choreography coordinators is avoided, but the choreographed service is generated as an indirect product of the pre-compiled behavior of the cooperating Web Services, without any support. In such cases, the exchange of messages between the cooperating WSs only depends on their interaction protocols, which are adapted to the choreography specification in order to enable the correct execution of the application.

Our approach is similar to the BECO system described in [15], which uses an event-based coordination paradigm. However, we relax the condition that enables the interaction of the Web Services in a choreography through the use of the Communication Interface component (CI). As long as the basic synchronization constraints are observed, in fact, the CI overcomes the problem of sending and receiving data as exactly specified in the WSDL interface.

[16] describes an interesting approach providing semi-automated support to identify and resolve mismatches between service interfaces and protocol, and for generating adapter specification. That approach offers service adapters mediating the interaction among two services with different interfaces, and it is implemented in Websphere environment. Our model addresses similar problems, but it is more general than that. In fact, while [16] concerns the mediation between two services, our proposal addresses the mediation among several services taking part in multi-party interaction: the choreography.

References

1. Peltz, C.: Web Services orchestration and choreography. Innovative Technology for Computing Professionals 36(10), 46–52 (2003)
2. W3C: Web Services Definition Language (2002), http://www.w3.org/TR/wsdl
3. Lu, R., Sadiq, S.: A survey of comparative business process modeling approaches. In: Abramowicz, W. (ed.) BIS 2007. LNCS, vol. 4439, pp. 82–94. Springer, Heidelberg (2007)
4. van der Aalst, W., van Hee, K.: Workflow Management - Models, Methods, and Systems. The MIT Press, Cambridge (2002)
5. Cabrera, F., Copeland, G., Freund, T., Klein, J., Langworthy, D., Orchard, D., Shewchuk, J., Storey, T.: Web Services Coordination (WS-Coordination) (2002), http://www-106.ibm.com/developerworks/library/ws-coor/
6. GigaSpaces: GigaSpaces SBA (2008), http://www.gigaspaces.com/pr_overview.html
7. Ahuja, S., Carriero, N., Gelernter, D.: Linda and friends. IEEE Computer 19(8), 26–34 (1986)
8. W3C: Web Services Choreography Description Language version 1.0 (2005), http://www.w3.org/TR/ws-cdl-10/
9. W3C: Web Services Conversation Language (WSCL) (2002), http://www.w3.org/TR/wscl10
10. Arkin, A., Askary, S., Fordin, S., Jekeli, W., Kawaguchi, K., Orchard, D., Pogliani, S., Riemer, K., Struble, S., Takacsi-Nagy, P., Trickovic, I., Zimek, S.: Web Service Choreography Interface 1.0 (2002), http://ifr.sap.com/wsci/specification/wsci-specp10.html
11. Ardissono, L., Petrone, G., Segnan, M.: A conversational approach to the interaction with Web Services. Computational Intelligence 20(4), 693–709 (2004)
12. DERI International: Web Service Modeling Ontology (2005), http://www.w3.org/Submission/WSMO/
13. DERI International: Web Service Modelling eXecution environment (2006), http://www.wsmx.org/
14. Benatallah, B., Casati, F., Grigori, D., Nezhad, H.M., Toumani, F.: Developing adapters for Web Services integration. In: Pastor, Ó., Falcão e Cunha, J. (eds.) CAiSE 2005. LNCS, vol. 3520, pp. 415–429. Springer, Heidelberg (2005)
15. Snoeck, M., Lemahieu, W., Goethals, F., Dedene, G., Vandenbulcke, J.: Events as atomic contracts for component integration. Data knowledge & knowledge engineering 51, 81–107 (2004)
16. Motahari Nezhad, H., Benatallah, B., Martens, A., Curbera, F., Casati, F.: Semi-automated adaptation of service interactions. In: Proc. of 16th Int. World Wide Web Conference (WWW 2007), Banff, CA, pp. 993–1002 (2007)

A Matchmaking Architecture to Support Innovation by Fostering Supply and Demand of Venture Capital

Mario Benassi[1], Tommaso Di Noia[2], and Alessandro Marino[3,4]

[1] Department of Economics Business and Statisitcs, University of Milan, Italy
mbenassi@dico.unimi.it
[2] SisInfLab, Politecnico di Bari, Via Re David, 200, Bari, Italy
t.dinoia@poliba.it
[3] IEGI, Bocconi University, Milan, Italy
[4] The Wharton School, University of Pennsylvania, Philadelphia, PA, USA
amarino@wharton.upenn.edu

Abstract. This paper proposes a web-based matchmaking engine aimed to overcome weaknesses of online directories – which typically includes only a list of supply features – by generating valuable information for users. Although applicable in several domains, the matchmaking engine has been developed for the venture capital industry where entrepreneurs seeking financing search for investors who, in turn, scout entrepreneurs on the basis of specific investment criteria. The match-making engine ultimate goal is to allow users to save time and money in their reciprocal search activity by generating analytic information on project compatibility measures.

1 Introduction

Venture capitalists (VC) are specialized companies financing start-ups and high prospective firms in exchange of minority or majority share. VCs aim is to obtain a capital gain by selling the shares after some years. VCs face the risk of losing their money, if the start-up does not take off, but have a chance of getting incredible rewards if the start-up succeeds. VCs are more than wise money lenders, as they provide start-ups with additional services such as advice, consulting and networking. Competition among VCs is increasingly severe in USA and Western Europe, for the market topped maturity and demand for financing slowed down after the 2001 bubble. As a consequence, well-timed access to information is a must and ability to identify attractive investments opportunities can make the difference . Financing start-ups is not a one-shot business: to satisfy their investors, VCs must rely on a steady flow of attractive project proposals that meet the requirements of their investment strategy . VCs do not search the market in a blind way, but look for projects with specific characteristics: a VC may prefer project in the software industry with a minimum investment threshold, while another may go for investments in smaller start-ups in other industries. Deal origination is the stage in which a VC scans the market looking for projects that match its specific requirements. In this stage, also entrepreneurs (ENT) scan the market: they explore financing opportunities for their start-ups with the goal of striking the best deal according to their preferences (e.g. money-equity ratio; rounds of financing: timing of

G. Psaila and R. Wagner (Eds.): EC-Web 2008, LNCS 5183, pp. 61–70, 2008.

exit, etc). Searching for a VC may become problematic and time consuming with possible negative impact on markets window exploitation for two main reasons. First, as financing has become a global international market, an appropriate VC can be located anywhere in the world. Distance may severely limit start-ups to search for appropriate VCs and vice-versa, as gathering complete information from both sides become problematic. Second, connection between start-ups and VCs is often indirect: ENTs may contact VCs through informants like consultants, professionals or business angels. On the other hand, VCs, receive proposals through a third party –such as investors, advisors and investors in the venture capital fund– who might have an economic incentive to sponsor promising projects and whose interests might be aligned with those of VCs. Contact through a third party may be beneficial, as it economizes on control costs and fuel trust, but it is not always possible. Should this be the case, ENTs directly search for VCs. The process is simple: public source provide basic information for an institutional contact, followed by a business-plan examination and eventually a due diligence. As a result, most prominent VCs tend to be overloaded with business ideas that lack referrals, while less prominent VCs and other investors will remain "invisible" to ENTs. Due to this mismatch entrepreneurs consume time and money for locating an appropriate VC. Paradoxically, some promising proposals will not find financing because they remain unknown. On the other hand VCs consume time and money to process un-referred proposals. Entrepreneurs should find appropriate VCs easier, whilst VCs should spare their time in low-value added activities. Several projects like Technet in US, Nordic Venture Network, Gate2Growth and Venture Route Projects in Europe moved in this direction. The idea is to act as infomediaries in the VC business[1].

2   Matchmaking Services

Infomediaries provide a variety of services – ranging from distributing publications describing projects seeking financing to selected investors, to organizing forums for firms and investors. Their services are similar to the ones available in an e-marketplace [2]. There, each transaction can be divided in three different steps [4]: **discovery, negotiation** and **execution**. In the first step, computer matching provides to the buyer a ranked list of offers based on her preferences. In the second step, the e-marketplace assists both seller and buyer in order to maximize their expected profit. In the execution step the buyer and the seller exchange the good. In the scenario of venture capitalists, matchmaking is not a one-shot process. It can be fractioned in further three separate phases: search for potential investors (classical matchmaking); submission of an application filled by the entrepreneur to an investor "found" during the previuos step; exchange both of information on the company and of a summary of the business. Very often, matchmaking is limited to Yellow-Page like service, namely a directory-based service where ENTs can look for a VC with specific features without having to know its name. Here the selection criteria are provided entirely by the potential entrepreneur. For VC scenarios, it is easy to identify two main limits in these services: firstly, they provide static information, preferences for VC investments preferences are set once and they

---

[1] http://tech-net.sba.gov, http://www.pminance.it, http://www.nordicventure.net,
  http://www.gate2growth.com, http://www.ventureroute.com

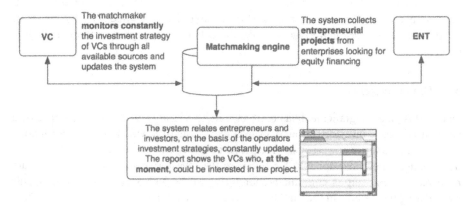

Fig. 1. The matchmaking activities

are not updated on a continuous basis; secondly, they do not offer to ENTs a valuable ranking of VCs and do not allow a focused effort in the initial stage, thus leaving the entrepreneur dealing with a list of names equally relevant.

A matchmaker (MM) has to allow an entrepreneur proposing a project to identify a VC whose investment strategy meets its characteristics and to what extent (see Figure 1). Specifically, a matchmaking process should comply with the following requirements:

(A) An effective use of sources for VC's investment criteria, selection and updating;
(B) A customized and clarifying reporting for ENTs.

VCs' investment strategies are not static, they can decide to change their investment target. In order to monitor these changes, the matchmaking system has to keep the dataset updated and reliable as much as possible, thus minimizing missing data. On the other hand, ENTs should obtain a valuable set of information from the matching system. Entrepreneurs would like to have a short list of VCs interested in the proposal they submitted, and understand why a VC ought to be more appropriate than another. Acting in such a way, a matchmaking system could be beneficial both to VCs and to ENTs. In fact, venture capitalists will receive only targeted projects, and entrepreneurs will be more likely to find the right investors. Both VCs and ENT would save time and reduce costs.

3 Rating-Based Matchmaking for Venture Capitalists

The goal of the rating-based matching process is to identify VCs that best fit a given ENT project. A key requirement for the whole process is to have a detailed set of data both for the entrepreneur and for the VC as well as a detailed report for ENT. Data collection involves the definition of new entries in the reference database, as well as a continuous updating of data previously retrieved. Once all the needed data have been collected then the matchmaking phase can be exectued. Matchmaking is computed in two steps: a data setting step and a matching step. Finally, in the very last phase, matching results are provided together with a description of the selected VCs by means of a reporting service. Hereafter, we describe the data collection phase and the matching process.

3.1 Data Collection

During this phase, significant data about VCs and ENTs are collected. First step is data entry of all active investors through a careful scanning of a wide range of information sources, such as investor associations, press, business and corporate web sites. In order to enter a new investor into the database the investors company contact data are retrieved. Second step is continuous monitoring of the VCs investment requirements by updating the data from the public resources collected in the industry mapping step, and by a direct contact with the investors. Data collected in the process have different origins, and therefore, a different value, such as the following scale: business web sites (low value), investors associations, Investors corporate web sites, press review, direct contact (highest value level)[2]. Highly valuable data require an updating through a direct contact with the investor. Existing literature on the venture capitalists criteria selection to assess entrepreneurial opportunities largely converges to a subset of relevant variables such as: (a) Life cycle stage; (b) Specialization; (c) Industry; (d) Amount of investment; (e) Geographical location; (f) Target level of profitability; (g) Management quality and composition; (h) target revenue; (i) Additional characteristics[5,6,8,3,9,1]. For each discrete option associated to all variables, further information has to be collected. The overall aim is to build a measure of the degree corresponding to how well each class of options fits an investor strategy. Such index is influenced by information collected from the investor, such as: (a) Investor's preferences (e.g. a specific industry or a specific technology); (b) Investor's track record and experience: that is, an evaluation of past activities; (c) Preferences directly expressed by the investor; (d) Number and type of operations carried out. Conversely, request of matchmaking by ENT implies gathering information on the project. This may happen according a procedure similar to the one listed above, namely with a multiple choice selection based on corresponding variables and options presented to an investor. In other words, the matchmaker extracts required options for each variable of the investment requirement that matches project characteristics. Furthermore, the matchmaker ranks related degree of option reliability by assigning a specific index based on further investigation on the entrepreneurial project. In the next sections we will provide some examples of the above procedures.

3.2 Matchmaking

The goal of the matchmaking algorithm is to find a set of relevant and potentially interested investors for a given project. The goal is achieved by means of an operator $match(e)$ that, given and ENT record e, returns a set $VC = \{vc_1, vc_2, \ldots, vc_n\}$ of investor references and data. Each investor record vc_i is also associated with a rating

---

[2] For the sake of conciseness we do not detail this step. The interested reader may refer to www.ventureroute.com

factor R representing the relevance measure of the investor with respect to the project. The match operator matches the investor's preferences with the features of each "investable" project and computes a score for each project. The higher the score, the more the "investable" project fits the investor's preferences. After the data setting phase, the score is combined with the measure of reliability associated with the given project, in order to select the most relevant investor for it (matching phase). The whole process is represented by Algorithm 1 (see also Figure 2).

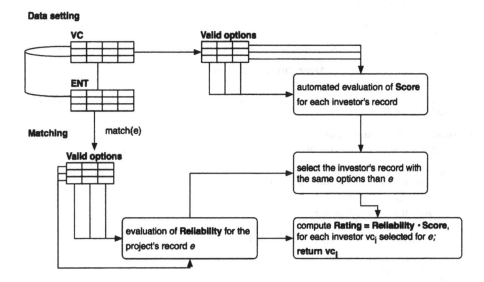

Fig. 2. The matchmaking process

Data Setting. As discussed above, the first phase of the algorithm process is the data setting. In this phase, for each investor record vc_i, we build a vector of preference on potential projects, which the investor could be interested in. This is done by selecting all the valid options that have been chosen by the investor and by returning the vector of the variables associated with them, according to Algorithm 1.

Given the records associated with each investor as returned by the data setting procedure, we calculate a score for each record that is a measure of the potential interest of an investor in a project. The following procedure, building on the previously discussed criteria to match projects' characteristics with investors' preferences, returns a score of interest of each investor, in a given project:.

where: (a) α is a coefficient of the relevance associated to each variable in each investor statement; (b) x represents the level of correspondence between the selected options and the investor activity; (c) $Molt$ is an indicator of the investor's level on an option and it is attributed automatically depending on the number of options selected; (d) y is the ratio variable/database and gives a higher score to investors with a strategy more focused in comparison to the industry; (e) finally $source$ is a measure of the reliability

---

Algorithm: *vc_data_setting*

Input: vc_i an investor record
Output: *Records_vci* an array of valid options for vc_i

begin
    $Records\_vci = ()$;
    foreach $valid\_option \in vc_i$ **do**
        Build a vector v containing the variables associated with *valid_option*;
        $Records\_vci = (Records\_vci, v)$;
    end
    return *Records_vci*;
end

Algorithm 1. *vc_data_setting*

Algorithm:*score*

Input: vc_i an investor record
Output: *Scores_vci* an array of scores associated to valid options of vc_i

begin
    $Records\_vci = vc\_data\_setting(vci)$;
    $Scores\_vci = ()$;
    foreach $r \in Records\_vci$ **do**
        $score = \alpha \cdot [(x \cdot Molt + y) \cdot source]$;
        $Scores\_vci = (Scores\_vci, \langle r, score \rangle)$;
    end
    return *Scores_vci*;
end

Algorithm 2. *score*

of the source which the information derives from. All the algorithm factors (i.e. from (a) to (e) values) are derived from information collected in the data collection phase [3]. However it is important to point out that each match maker can define its own method to establish the relative importance of each variable's options for each VC. Nevertheless, according to the efficient market hypotheses [7], it is reasonable to expect a general consensus among the matchmakers methods as the availability of relevant information tends to overlap.

The Algorithm. Once the data setting phase is completed, the matching algorithm is ready to compute the match degree for a given project e. The first step of matching process is similar to the investor's profiling, we select valid options in e describing the project and we associate a measure of its reliability –instead of the score computed for

---

[3] For example the value of the factor $x = (C1, C2, C3)$ can be obtained by dividing the considered range into three parts. One third of the score can be attributed on the basis of the number and type of operations carried out ($C1$). One third is attributed on the basis of the investor's sentiments regarding the company strategic vision ($C2$). Finally, the other third is attributed on the basis of match-maker's perceptions regarding investor track record and experience ($C3$).

Algorithm:*env_data_setting*

Input: a given project e

Output: $Records\_e$ a set of valid options and associated reliability

begin
    $Records\_e = ()$;
    foreach $valid\_option \in e$ **do**
        Build a vector v of the variables associated with $valid\_option$;
        Ask the designer to set the $reliability(v)$ associated with v in $[0, 1]$;
        $Records\_e = (Records\_e, \langle v, reliability(v) \rangle)$;
    end
    return $Records\_e$;
end

Algorithm 3. *env_data_setting*

Algorithm:*match*

Input: $e = \langle v, reliability(v) \rangle$

Output: VC

begin
    $VC = ()$;
    $ENT = ent\_data\_setting(e)$;
    $Reliability = reliability(v)$;
    $Candidate\_vc = \{vc_i \mid vc_i \; is \; featured \; by \; the \; same \; options \; than \; e\}$;
    foreach $vc_i \in Candidate\_vc$ **do**
        $Score = get\_score(vc_i)$;
        $Rating = Reliability \cdot Score$;
        $VC = (VC, \langle vc_i, Rating \rangle)$;
    end
    return VC;
end

Algorithm 4. *match*

the investor– with each option of each record of e. The main difference is that the reliability is chosen by the matching designer in a range $[0, 1]$ according to the information collected from ENT. The interactive procedure is defined in Algorithm 3. The matchmaking process uses the *env_data_setting* and the *vc_data_setting* procedures to get the reliability and the score for each project and for each investor. Reliability and score return a set of suitable investors in a given project according to Algorithm 4. Summing up, once the data setting on investors is computed, the matchmaking procedure gets the data setting and the reliability of the project as well as a set of suitable investors. Suitable investors have the same options listed in the project by ENT. For each suitable investor, the project reliability index is combined with the investors score to get a rating factor, thus ordering the investors suitable to the project. The main idea behind the matchmaking operation is to contextualize a general evaluation of the VCs' investment preferences, represented by the score, with respect to the specific features of the entrepreneurial project, represented by the reliability index. In such a way, the matchmaking operation generates an ordered set of investors that are potentially interested in the specific project.

4 A Simulation of the Process

In this section we provide a step by step example of the matchmaking process described in the previous section. We use a toy context to better explain the process. For the sake of simplicity let us assume that:

- Three VCs have been profiled: VC1, VC2 and VC3;
- ENT makes a matchmaking request;
- The investment strategies of VCs include a multiple choice selection among the following variables: (a) Life cycle phase (related options: (a.1) start-up; (a.2) Small-Medium-Enterprise (SME)); (b) Investment sectors (related options: (b.1) Biotech; (b.2) Automotive components;
- The measure used to rank the VCs preference degree for each option is given by the simplified expression: $score = \alpha \cdot x \cdot source$ (with $Molt = 1$ and $y = 0$) where:
  - α are variables representing preference scores directly suggested by VCs. These variables range in $[0,1]$;
  - x shows correspondence between the options and the investor activity. It has a discrete range $[0, \ldots, 100]$;
  - $source$, represents the measure of the source reliability. It is set to 1 if a VC provide that information directly or 0.5 if MM gets it from the web.

Accordingly, a data collection is first conducted by MM obtaining the data shown in Table 1. In the next step of the process a data set is built according to the procedures $vc\_data\_setting$, $score$. The results are summarized in Table 2[4]. To perform a match, we have to build a profile of the project. We use a matrix similar to the one used for VCs, thus selecting the options and their reliability coefficient. In our example we assume that ENT has a start-up project in the biotech industry. Accordingly $env\_data\_setting$ returns the results shown in Table 3.Finally $match$ procedure performs three tasks: (i)

Table 1. Examples of data collected for investor's files creation

Investor	Preference for (a) Life cycle phase	Preference for (b) Life cycle phase	x	Source	α for (a)	α for (b)
VC1	(a.1) start-up (a.2) SMEs		75	1	0.5	1
			100	0.5		
		(b.1) Biotech	100	0.5		
VC2	(a.1) start-up (b.1) Biotech		100	1	1	0.5
			100	1		
		(b.2) Automotive components	50	1		
VC1	(a.1) start-up (a.2) SMEs		30	0.5	0.3	0.7
			75	0.5		
		(b.2) Automotive components	100	1		

---

[4] We normalize scores as real numbers in $[0, \ldots, 100]$.

Table 2. Data set of investors' preferences

Investor	Variable (a)	Score (a)	Variable (b)	Score (b)
VC1 (a)	(a.1) start-up	37.5	(b.1) Biotech	25
VC1 (b)	(a.2) SMEs	25	(b.1) Biotech	25
VC2 (a)	(a.1) start-up	100	(b.1) Biotech	50
VC2 (b)	(a.1) start-up	100	(b.2) Automotive components	25
VC3 (a)	(a.1) start-up	4.5	(b.2) Automotive components	70
VC3 (b)	(a.2) SMEs	11.25	(b.2) Automotive components	70

Table 3. Data set of ENT project profile

Variable (a)	Reliability index (a)	Variable (b)	Reliability index (b)
(a.1) start-up	0.75	(b.1) Biotech	100
(a.2) SMEs	0.25	(b.1) Biotech	100

Table 4. Results of the rating based matchmaking process

Rating (R)	Investor	Variable (a)	Option rating (a)	Variable (b)	Option rating (b)
125	VC2 (a)	(a.1) start-up	75	(b.1) Biotech	50
53.125	VC1 (a)	(a.1) start-up	28.125	(b.1) Biotech	25
31.25	VC1 (b)	(a.2) SMEs	6.25	(b.1) Biotech	25

matches ENT profile with the VCs data set; (ii) computes an option rating; (iii) sorts the resulting records in decreasing order by general ratings. In our example we obtain the report shown in Table 4.

5   Discussion and Conclusion

The process we described meets the requirements (A) and (B) of Section 2. Our ultimate goal was to propose a matchmaking engine able to overcome the basic contribute of simple online directories and to build a process where both MM and the automated system cooperate in benefiting both ENT and VCs. The example we report in Section 4 shows how the process saves time and reduces costs; by taking a look at the report, ENT can easily conclude that: (1) VC2 has the closest investment strategy for ENT features, VC1 even with lower ratings can be potentially interested, VC3 does not match; (2) Each option rating reveals the reasons of the similitude or differences between the projects features and the investment preferences. Thus the process provides information valuable for all: it allows ENT to find the most suitable VCs, and VCs to receive only targeted projects. Low scores on specific variables should discourage ENT to submit low rated projects, but could offer the opportunity of addressing unsolved issues, thus complying with the requirement (B). Finally, the system allows MM to focus only on monitoring and updating information on VCs investment strategy from a pool of sources (with different degree of reliability). Therefore, the requirement (A) is satisfied.

The algorithm we propose has already been adopted for public and commercial use. In its earlier version it has been first tested and developed in 2002 by the Italian Consorzio Camerale; even if that version was not available on the web, it produced almost a hundred of reports and obtained positive feedback from the users. Later on it was further developed and redesigned as a web-based application within the project VentureRoute — a European Commission sponsored project— which involved a group of partners from different European countries. We believe that the algorithm and the matchmaking process could be highly beneficial both for entrepreneurs looking for financing and for venture capitalists looking for good investment opportunities, but also for other parties. As we mentioned, financing start-up and SME is not always a dyadic business, as it involves other parties such as banks, consultants, etc. A ranking service and a detailed report could be helpful also for these players and let them focus on the right partners: a bank might leverage its contacts with a specific VC, instead of searching the market in a blind way. Eventually, the matchmaking process we propose could also be beneficial for institutions and public policies, encouraging them to devise appropriate initiatives. The matchmaking process could also incentive brokers and third parties. Brokers and third parties play a crucial role in favoring start-ups, still their role may be severely limited by spatial and geographical adverse conditions. Acting as a broker in a district where VCs abound is easier than in districts barely populated.

Acknowledgements. Tommaso Di Noia acknowledges partial support of EU-FP-6-IST-STREP-026896 TOWL project.

References

1. Zopounidis, C.: Venture capital modelling: Evaluation Criteria for the appraisal of investments. The Financer ACMT 1(2), 54–64 (1994)
2. Hoffner, Y., Facciorusso, C., Field, S., Schade, A.: Distribution issues in the design and implementation of a virtual market place. Comput. Netw. 32(6), 717–730 (2000)
3. Siskos, J., Zipounidis, C.: A Model of Venture Capitalist Investment Activity. J. of business venturing 30(9), 1051–1066 (1987)
4. MacKie-Mason, J.K., Wellman, M.P.: Automated markets and trading agents. In: Handbook of Computational Economics. North-Holland, Amsterdam (2006)
5. Macmillan, I.C., Siegel, R., Subbanarasimha, N.P.: Criteria used by venture capitalists to evaluate new venture proposals. Journal of Business Venturing 1(1), 119–128 (1985)
6. Macmillan, I.C., Zemann, L., Subbanarasimha, P.N.: Criteria distinguishing successful from unsuccessful ventures in the venture screening process. Journal of Business Venturing 2(2), 123–137 (1987)
7. Malkiel, B.G.: The efficient market hypothesis and its critics. Journal of Economic Perspectives 17(1), 59–82 (2003)
8. Siskos, J., Zopounidis, C.: The evaluation criteria of the venture capital investment activity: An interactive assessment. European Journal of Operational Research 31(3), 304–313 (1987)
9. Wells, W.A.: Venture Capital Decision-making. PhD thesis, Carnegie-Mellon (1974)

Using Expressive Dialogues and Gradient Information to Improve Trade-Offs in Bilateral Negotiations

Ivan Marsa-Maestre, Miguel A. Lopez-Carmona, Juan R. Velasco, and Bernardo Alarcos

Departamento de Automatica
Universidad de Alcala
Alcala de Henares, Spain
{ivan.marsa,miguelangel.lopez,juanramon.velasco,bernardo.alarcos}@uah.es

Abstract. A bilateral negotiation may be seen as an interaction between two parties with the goal of reaching an agreement over a given range of issues which usually involves solving a conflict of interests between the parties involved. In our previous work, we address the problem of automatic bilateral negotiation by using fuzzy constraints as a mean to express participant's preferences, focusing in purchase negotiation scenarios. Other research works have used similarity criteria to perform trade-offs in bilateral bargaining scenarios, without any expressive mechanisms between participants. In this paper, we combine our expressive approach with the traditional positional bargaining schema. In particular, we explore the possibility of using the derivatives of each agent's valuation function to issue direction requests to narrow the solution search space of its counterpart, thus improving the effectiveness and efficiency of the negotiation over traditional positional approaches.

1 Introduction

Automated negotiation is an important challenge in the electronic commerce community which has been covered from different areas such as game theory [1], distributed artificial intelligence [2] and economics [3]. In particular, this paper covers multi-issue bilateral negotiations, which involve a bargaining process between two parties or agents (a *player* and an *opponent*), which exchange proposals (*contracts*) in order to reach an agreement over a given range of issues. In these scenarios, the negotiation process and outcome is determined by the participant's utility functions and the impact that the different issues under negotiation have over the utility that each proposed contract yields to the negotiating parties. If the impact of the issues under negotiation over the utility function is different for each agent (that is, some issues are more important for the player than for the opponent and vice versa), the issues may be traded-off against one another, increasing the social welfare of the deal [4]. An approach to deal with such scenarios is described in [5], but the negotiation framework proposed there is mainly based on positional bargaining which, according to

G. Psaila and R. Wagner (Eds.): EC-Web 2008, LNCS 5183, pp. 71–80, 2008.

negotiation theorists [4], is a mechanism which may be improved by knowing information about the other negotiating parties. Taking this into account, we propose to take advantage of the expressive power of dialogue games to lead the trade-off algorithm to reach a more satisfying solution for both parties in a more efficient manner. In this paper a mechanism which uses information about the derivatives of the agent's valuation functions within the trade-off algorithm is proposed. By performing experiments comparing the original trade-off algorithm to our mechanisms we show how this expressive approach may provide benefits in terms of performance and optimality over previous works.

The rest of the paper is organized as follows. Section 2 recalls the most relevant previous work our research is related to. Section 3 describes our approach for symmetric bilateral negotiation using expressive dialogues and introducing derivatives in the trade-off algorithm. The experimental evaluation is provided in Section 4. The last section summarizes our main contributions and sheds light on some future research.

2   Similarity-Based Negotiation Trade-Offs

In [5], an algorithm for carrying out trade-offs in automated negotiations is proposed. In our previous work we defined a framework where agreements were pursued through widening the space of acceptable solutions of buyer and seller until an intersection appears (and thus involving a certain utility loss for one or both participants). In contrast, in [5] the authors assume that such intersection already exists and focuses on finding the intersection through an iterated hill-climbing search in a landscape of possible contracts. Contracts are defined as sets of values for the different issues which are being negotiated, and agents' utilities for a given contract are computed using a weighted sum of monotonically increasing or decreasing scoring functions for each issue. Also, the concept of *iso-curve* is defined as the curve comprising the solutions which yield a given utility for a given agent. The interaction protocol is a positional bargaining, that is, only specific solutions to the negotiation problem are exchanged between the agents. Once both agents participating in the negotiation have proposed an initial solution, solutions proposed in the subsequent steps of the negotiation are points lying in the same iso-curve that the last own proposal while maximizing the similarity to the opponent's last offering. The search for the next proposal to make is performed by successively generating random contracts which lay closer to the *iso-curve* and selecting the more similar contract to the opponent's proposal. The algorithm terminates at each step in the negotiation when the last selected contract lies in the *iso-curve*.

3   Expressiveness in Symmetric Bilateral Negotiation Dialogues

Though both the research work mentioned in Section 2 and the framework we proposed in [6] address automated multi-attribute bilateral negotiation in significantly different ways, we believe that these two approaches can be combined, taking advantage of the expressive power of dialogue games to lead the trade-off algorithm to reach a more satisfying solution for both parties in a more efficient manner. In this work we propose a derivative-based mechanism to take into account agent expressiveness during trade-offs. Furthermore, we define a dialogue to allow agents to negotiate expressing their preferences, searching for trade-offs or making concessions as necessary. While agent willingness to truthfully reveal preference information can be questionable, we can assume a risk-averse scenario, where not reaching an agreement in a given time has a negative payoff for the negotiating agents.

3.1   Expressiveness and the Trade-Off Algorithm

In [7] we evaluated the effect of the agents' attitudes in terms of expressiveness and receptiveness in asymmetric automated purchase negotiations. For the seller agent, an expressiveness parameter controls whether the seller agent expresses its preferences for a specific relaxation of the previous buyer's demands, while a receptivity parameter modulates the seller's attitude regarding the buyer's purchase requirements. For the buyer agent, an expressiveness parameter controls the use of purchase requirement valuations, while a receptivity parameter modulates the buyer's attitude regarding a relaxation requirement received from a seller agent. Here we propose to apply analogous concepts to symmetric negotiation scenarios and the trade-off algorithm.

Following a similar notation to the one used in the related works, we define the issues under negotiation as a finite set of variables $x = \{x_i | i = 1, ..., n\}$, and a contract (or a possible solution to the negotiation problem) as a vector $s = \{x_i^s | i = 1, ..., n\}$ defined by the issues' values. The *overall (or global) utility* of a potential solution s is $V(s) = \oplus \{V_i(x_i^s) | i = 1, ..., n\}$, where $\oplus$ is an aggregation from $[0, 1]^n$ to $[0, 1]$, and $V_i(x_i)$ is the agent scoring function for the issue x_i. In this work we are dealing with symmetric bilateral negotiations, involving two agents, named *player* and *opponent*. Since the algorithm used is symmetric, all discussion can be done from the viewpoint of the player without losing generality.

For the sake of simplicity, we restrict ourselves to weighted additive aggregation functions and independent scoring functions for each issue in the negotiation. That is, the overall utility of a potential solution s for an agent j is $V^j(s) = \sum_{1 \leq i \leq n} \omega_i^j V_i^j(x_i^s)$, where $W^j = \{\omega_i^j | i = 1, ..., n\}$ models the importance that agent j assigns to each decision variable i under negotiation as a weight ω_i^j.

Within this framework, we define a derivative-based mechanism to introduce expressiveness in the trade-off algorithm and a negotiation dialogue intended to take advantage of such approach.

3.2   Using Derivatives within the Trade-Off Algorithm

The trade-off algorithm [8] performs an iterated hill-climbing search over the solution space. This is done by starting at the opponent's last proposal, y, and moving towards the iso-curve associated with the utility of the player's last proposal. The algorithm performs a total of S steps, and at each step it generates N children contracts which are closer to the iso-curve than the ones in the previous step by an amount of utility E. From all the children, the most *similar* to the opponent last proposal is selected as the starting point for the next step. In this way, at each step of the algorithm the contract under consideration is closer to the utility of the player's last offer, while maximizing similarity to the opponent's last offer. The algorithm stops when the contract under consideration lies in the same iso-curve as he player's last proposal (i.e it has the same utility).

The algorithm generates children by splitting the gain in utility randomly among the set of issues under negotiation. For each issue i, the algorithm assign an utility increase for this issue $r_i = min(random(E_i), \frac{E-E_n}{\omega_i})$, where E_i is the maximum possible gain for the issue x_i at this step and $random(E_i)$ generates a random number between 0 and E_i. The utility increase is limited by $\frac{E-E_n}{\omega_i}$, which is computed as the difference between the target increase in utility at this step E and the already accumulated utility gain at this step E_n. This limits the final gain of the step to E. Since the generation of children at each step of the hill climbing process is random, a mechanism which may be used to increase the effectiveness and efficiency of the search for solutions is to perform a more *directed* hill-climbing, that is, to generate the children at each step in the direction that causes the least utility loss to the opponent while increasing the agent's own utility.

To this end, we allow an agent to generate a specific *direction request*, intended to influence the hill-climbing path followed to generate the intermediate solutions for the different steps of the trade-off algorithm. A *direction request* is defined as a vector $d_{req} = \{d_i | i = 1, ..., n\}$, where d_i is computed by normalizing the partial derivatives $\frac{\partial V(s)}{\partial x_i}$ of the global utility function of the agent issuing the request at the point defined by its opponent proposal. What we propose is using this information to modulate the random utility gain splitting computed in each iteration of the trade-off algorithm. Since the partial derivatives of the scoring functions express how an agent's utility varies with the variation of each individual issue, this information may be used to weigh the utility increase for each issue at each step, so that the utility increase is performed mainly over the attributes that impact less the other agent's utility. The point in the algorithm to perform this modulation is when the algorithm assign an utility increase r_i for each issue x_i. The utility increase is defined as $r_i = min(random(\frac{E_i}{d_i}), E_i, \frac{E-E_n}{\omega_i})$, thus assigning more utility gain to those issues where the partial derivatives $\frac{\partial V(s)}{\partial x_i}$ express a lesser impact over the opponent's utility and vice versa. Our hypothesis is that using derivatives in this way within the algorithm will direct the hill-climbing process to solutions that, while keeping the agent's utility constant, have a lesser impact over the opponent's utility, thus improving the outcome of the trade-off algorithm in terms of player and opponent's utility. Furthermore, by restricting the hill-climbing path to a direction known to provide more satisfying solutions, less children will be needed to achieve a certain outcome, again improving algorithm efficiency.

3.3   Negotiation Dialogue

The negotiation protocol follows a dialogue game approach analogous to the one described in [6]. Due to space reasons, only the negotiation stage of the dialogue is described here. At the first step in the negotiation, both agents propose initial solutions to the problem using the locution *L1: Propose_ solution*. Any of the agents may now either apply the trade-off algorithm over the current solutions *or* express their preferences through locution *L3: Request_ direction*, which issues a direction request. The expressiveness and receptiveness parameters of each agent govern the probabilities of use of each one of the decision mechanisms and locutions. For moderately expressive and receptive agents, a typical negotiation dialogue is as follows:

1. The initiator issues a first solution proposal, and the responder issues back another solution.
2. The initiator computes next proposal using the trade-off algorithm and sends it to its opponent.
3. The responder computes next proposal using the trade-off algorithm and sends it to the initiator.
4. Steps 2 and 3 are repeated until a deadlock as defined in [9] is detected (i.e. an agent detects that its opponent's proposal no longer increase its own utility). The agent detecting the deadlock sends a direction request to its opponent indicating the normalized partial derivatives of its valuation function at the point defined by the last received proposal.
5. The opponent receiving the direction request computes next proposal using the information received to modulate the random utility gain splitting computed in each iteration of the trade-off algorithm. The new proposal is sent to the agent detecting the deadlock.
6. If the new solution received ends the deadlock, agents proceed to step 4. Otherwise, the agent may repeat the algorithm for another iso-curve (thus conceding to lose a certain degree of utility) *or* send a relax requirement to its opponent (requesting her to lose a certain degree of utility).

4   Experimental Analysis

Our experiment plan is designed to determine whether the proposed mechanism provides an improvement to the efficiency and optimality of the negotiation process over the previous work described in Section 2. To this end, we have reproduced the experiments performed in [5], comparing the results of the original trade-off algorithms with the results obtained applying the derivative-based approach. As in this work is suggested, *single offer experiments* have been performed to evaluate the contribution of the proposed mechanism to the trade-off algorithm, and *meta-strategy experiments* have been devised to evaluate the contribution of the use of expressiveness to the overall negotiation process.

4.1   Experimental Settings

In single-offer experiments, the experimental procedure consists of inputting two contracts (representing the agent's initial utterances) into the algorithm and observing the execution trace of the algorithm for *one* offer from the *player* to the *opponent* (i.e. observing how the algorithm climbs from the opponent's proposal to a new proposal which has the same utility as the player's initial proposal in S steps).

The purpose of the meta-strategy experiments is to evaluate the outcome and dynamics of the negotiation when agents use either a trade-off mechanism or a concession mechanism (that is, a mechanism involving utility loss for the agent at each iteration) or a combination of the two in the course of the negotiation. To do this, the execution trace of the negotiation process is observed, recording the interchange of proposals between agents during the negotiation when using different meta-strategies. The meta-strategies considered have been limited to the set $\{smart, serial, random\}$. The smart strategy pursues a tradeoff until a deadlock occurs, then switching to a concession mechanism with an utility concession of 0.1 at each new proposal. A serial strategy involves alternating between the trade-off and responsive mechanisms, and a random strategy randomly choses between the two at each step in the negotiation.

Both types of experiments involve the negotiation between two agents, named *player* and *opponent*, over four quantitative issues, using two contracts as the starting point of the negotiation. Contracts are chosen so that they give a high utility to the proposer and a low utility to its counterpart. As in Faratin's work [5], we restrict ourselves to an additive and monotonically increasing or decreasing scoring system, using the same utility functions and the same criteria for computing the similarity. The importance weight vectors of the agents, used to compute the global utility function for each agent, are fixed throughout the negotiation: $W_{player} = [0.15, 0.25, 0.1, 0.5]$ and $W_{opponent} = [0.5, 0.15, 0.1, 0.25]$.

To evaluate the contribution of the proposed mechanisms to the algorithm in terms of effectiveness, we have performed the experiments for the best case described in [5], using $S = 40$ as the number of steps to reach the iso-curve and $N = 100$ as the number of children generated at each step, and assuming perfect knowledge to compute similarity (that is, the weights of each agent are known to the other). The perfect knowledge assumption also implies knowledge of the derivatives, since for a linear additive scoring system, the opponent weights $W_{opponents}$ *equal* the partial derivatives of the valuation function. As observed in [9], the order in which the different issues are processed by the trade-off algorithm greatly impacts the final outcome. Taking into account this, we have repeated the experiments for different issue orderings. Finally, to evaluate the contribution of the different mechanisms to the algorithm in terms of efficiency, we have performed a set of experiments varying the number of children N, in order to test our hypothesis that fewer children are needed to achieve the same result when using derivatives within the trade-off algorithm.

4.2 Experimental Results

Figure 1 shows the results of the single-offer experiments. Each graphic presents the box-plot for the final outcomes of 100 runs of the trade-off algorithm. The horizontal axis represents the algorithm under evaluation: the original trade-off algorithm, the derivative-based expressive approach, and of a random reference algorithm without using expressiveness or similarity. In the vertical axis we have represented the median and 25th and 75th percentiles of the opponents' utility (player's utility does not change during a single run of the trade-off algorithm). We can see that there is a significant improvement of the utility of the final outcome for the opponent, and that the improvement is more significant for some orderings, yielding utility gains of nearly 80% over the approach in [8]. From these results we can conclude that the use of derivatives makes the trade-off algorithm more robust to the ordering of the issues.

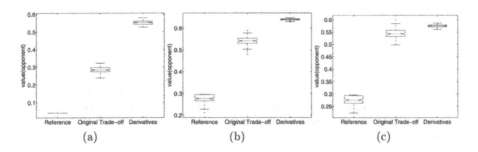

Fig. 1. Effect of the use of knowledge about derivatives ($W_{opponent}$) over the trade-off algorithm under different issue orderings: a) order [1 2 3 4], b) order [3 2 4 1], c) order [1 4 2 3]

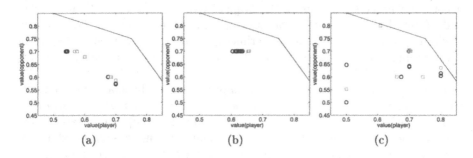

Fig. 2. Final outcomes of the negotiation process for meta-strategies running the trade-off algorithm with 100 children a) *smart* vs. *smart*, b) *serial* vs. *serial*, c) *random* vs. *random*

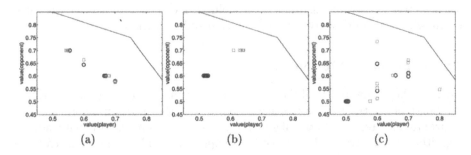

Fig. 3. Final outcomes of the negotiation process for meta-strategies running the trade-off algorithm with 50 children a) *smart* vs. *smart*, b) *serial* vs. *serial*, c) *random* vs. *random*

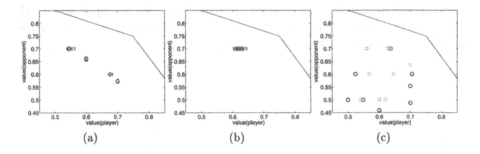

Fig. 4. Comparison of the results obtained with the inexpressive approach for 100 children and the results obtained with the expressive approach for 50 children for the same meta-strategy pairings: a) *smart* vs. *smart*, b) *serial* vs. *serial*, c) *random* vs. *random*

Figure 2 shows the results of the meta-strategy experiments when the trade-off algorithm is run with 100 children. Each graphic shows the results of 10 runs of the experiment. The x-axis and y-axis represent, respectively, the *player* and *opponent* utilities. For each run we have represented the final outcome of the negotiation process when using the original trade-off algorithm (black circles) and of our derivative-based expressive approach (grey squares). For the sake of clarity, no random reference experiment has been represented. A black line joining represents the Pareto-optimal line, computed using the weighted utility functions method [4,10].

We can see that there is a significant improvement of the joint utility of the final outcome for the negotiations where both agents follow *smart* meta-strategies and a slight improvement in the case of the [*serial*, *serial*] meta-strategy pairing. In a *random* vs. *random* scenario, the final outcomes are scattered and no significant conclusions can be extracted in favour of one or other approach.

Figure 3, where the results of the meta-strategy experiments when the trade-off algorithm is run with 50 children are represented, show how the improvements observed in some combinations of metastrategies become more evident when the number of children is reduced. Again, the results obtained for a *random* vs. *random* scenario are unconclusive because of the scattering of the final outcomes.

Finally, Figure 4 compares derivative-based approaches to the original trade-off algorithm in terms of efficiency, showing the metaestrategy results when running the trade-off algorithm with 100 children for the inexpressive approach and with 50 children for the derivative-based expressive approach. In both cases, we achieve fairly similar results in terms of joint gain. Therefore, we can conclude that using derivatives within the trade-off algorithm can yield to a significant reduction of the number of children needed for a given outcome, thus increasing the efficiency of the negotiation process.

5 Conclusions and Future Work

There are vastly different research directions regarding automatic bilateral negotiations covering different areas such as game theory, evolutionary computation and distributed artificial intelligence, many of them involving integrative negotiation mechanisms [11,12]. We have worked in the past with expressive dialogue games in asymmetric bilateral purchase negotiation scenarios [6]. In [5], a trade-off algorithm for more generic symmetric negotiation scenarios is proposed, which is mainly based on positional bargaining. In this paper, we combine our negotiation framework with the trade-off algorithm to show how the use of expressive dialogues may improve the efficiency and optimality of the negotiation process over traditional positional approaches. In particular, we have developed and evaluated a mechanism to improve the trade-off algorithm through expressiveness by using knowledge about the derivatives of the opponent's valuation functions to influence the direction in which new solutions are searched for. There are other research works that propose the use of derivatives [13] or constraints [14,15] to increase the joint gain in bilateral automatic negotiations. However, these approaches assume the existence of a trusted third party who acts as a mediator in the negotiation. Our proposal, in contrast, consider the interchange of information directly between the agents, so no mediator is involved.

Though the experiments have yielded satisfactory results, there is still plenty of research work to be done in this area. A more in-depth performance analysis of the algorithm is main priority for future work. An adaptation of the trade-off algorithm to work with discontinuous scoring functions and discrete solution sets should be made as well. Finally, we are interested in evaluating the performance of the different approaches with time-dependent scoring functions, to test the adequacy of the automated negotiation techniques for real-world, dynamic negotiation scenarios.

References

1. Rosenschein, J.S., Zlotkin, G.: Rules of Encounter. MIT Press, Cambridge (1994)
2. Faratin, P., Sierra, C., Jennings, N.R.: Negotiation decision functions for autonomous agents. Robotics and Autonomous Systems 24(3-4), 159–182 (1998)

3. Raiffa, H.: Lectures on negotiation analysis. PON Books, Harvard Law School (1996)
4. Raiffa, H.: The Art and Science of Negotiation. Harvard University Press (1982)
5. Faratin, P., Sierra, C., Jennings, N.: Using similarity criteria to make negotiation trade-offs. In: Proceedings of the 4th International Conference on Multi-Agent Systems, pp. 119–126 (2000)
6. Lopez-Carmona, M.A., Velasco, J.R.: A fuzzy constraint based model for automated purchase negotiations. In: Fasli, M., Shehory, O. (eds.) TADA/AMEC 2006. LNCS (LNAI), vol. 4452, pp. 234–247. Springer, Heidelberg (2007)
7. Lopez-Carmona, M.A., Velasco, J.R., Marsa-Maestre, I.: The agents' attitudes in fuzzy constraint based automated purchase negotiations. In: Burkhard, H.-D., Lindemann, G., Verbrugge, R., Varga, L.Z. (eds.) CEEMAS 2007. LNCS (LNAI), vol. 4696, pp. 246–255. Springer, Heidelberg (2007)
8. Faratin, P., Sierra, C., Jennings, N.R.: Using similarity criteria to make issue trade-offs in automated negotiations. Artificial Intelligence 142(2), 205–237 (2002)
9. Ros, R., Sierra, C.: A negotiation meta strategy combining trade-offs and concession moves. Autonomous Agents and Multi-Agent Systems 12(2), 163–181 (2006)
10. Ehtamo, H., Ketteunen, E., Hamalainen, R.P.: Searching for joint gains in multi-party negotiations. European Journal of Operational Research 1(30), 54–69 (2001)
11. Kersten, G.E., Noronha, S.J.: Rational agents, contract curves, and inefficient compromises. IEEE Transactions on Systems, Man and Cybernetics, Part A Systems and Humans 28(3), 326–338 (1998)
12. Klein, M., Faratin, P., Sayama, H., Bar-Yam, Y.: Protocols for negotiating complex contracts. IEEE Intelligent Systems 18(6), 32–38 (2003)
13. Vo, Q.B., Padgham, L., Cavedon, L.: Negotiating flexible agreements by combining distributive and integrative negotiation. Intelligent Decision Technologies 1(1-2), 33–47 (2007)
14. Ehtamo, H., Hamalainen, R.P., Heiskanen, P., Teich, J., Verkama, M., Zionts, S.: Generating pareto solutions in a two-party setting: constraint proposal methods. Management Science 45(12), 1697–1709 (1999)
15. Ito, T., Hattori, H., Klein, M.: Multi-issue negotiation protocol for agents: Exploring nonlinear utility spaces. In: Proceedings of the International Joint Conference on Artificial Intelligence, IJCAI 2007, pp. 1347–1352 (2007)

Towards an Interdisciplinary Framework for Automated Negotiation

Fernando Lopes[1], A.Q. Novais[1], and Helder Coelho[2]

[1] INETI, Dep. de Modelação e Simulação, Est. Paço Lumiar, 1649-038 Lisboa, Portugal
{fernando.lopes,augusto.novais}@ineti.pt
[2] Universidade de Lisboa, Dep. de Informática, Campo Grande, 1749-016 Lisboa, Portugal
hcoelho@di.fc.ul.pt

Abstract. Negotiation is an important and pervasive form of social interaction. The design of autonomous negotiating agents involves the consideration of insights from multiple relevant research areas to integrate different perspectives on negotiation. As a starting point for an interdisciplinary research effort, this paper presents a model that handles bilateral multi-issue negotiation, employs game-theoretic techniques to define equilibrium strategies for the bargaining game of alternating offers, and formalizes a set of negotiation strategies and tactics studied in the social sciences. Autonomous agents equipped with the model are currently being developed using the Jade framework. The agents are able to negotiate under both complete and incomplete information, thereby making the model in particular and the agents in general very compelling for automated negotiation.

1 Introduction

Autonomous negotiating agents representing individuals or organizations and capable of reaching mutually beneficial agreements are becoming increasingly important. Examples, to mention a few, include the business trend toward agent-based supply chain management, the pivotal role that electronic commerce is increasingly assuming in organizations, and the industrial trend toward virtual enterprises (see, e.g., [2,4]).

Artificial intelligence (AI) researchers have paid some attention to automated negotiation over the last years and a number of models have been proposed in the literature. These models can be classified into three main classes [7]: (i) game-theoretic models, (ii) heuristic models, and (iii) argumentation-based models. Game-theoretic models provide clear analysis of specific negotiation situations and precise results concerning the optimal strategies negotiators should choose, *i.e.*, the strategies that maximize negotiation outcome (see, e.g., [5,8]). Heuristic models provide general guidelines to assist negotiators and beneficial strategies for moving toward agreement, *i.e.*, strategies that lead to good (rather than optimal) outcomes (see, e.g., [3,6]). Typically, they are based on informal models of interaction and negotiation from the social sciences. Argumentation-based models allow negotiators to argue about their mental attitudes during the negotiation process. Thus, in addition to submitting proposals, negotiators can provide arguments either to justify their negotiation stance or to persuade other negotiators to change their negotiation stance (see, e.g., [16]).

G. Psaila and R. Wagner (Eds.): EC-Web 2008, LNCS 5183, pp. 81–91, 2008.

Game-theoretic models have some highly desirable properties, such as Pareto efficiency and the ability to guarantee convergence. However, most models make the following restrictive assumptions: (i) the agents are rational, (ii) the set of candidate solutions is fixed and known by all the agents, and (iii) each agent knows either the other agents' payoffs for all candidate solutions or the other agents' potential attitudes toward risk and expected-utility calculations.

Heuristic models exhibit the following desirable features: (i) they are based on realistic assumptions, and (ii) they make use of moderate computational resources to find acceptable solutions (according to the principles of bounded rationality [18]). However, most models lack a rigorous theoretical underpinning – they are essentially ad hoc in nature. Also, they often lead to outcomes that are sub-optimal. Finally, there is often no precise understanding of how and why they behave the way they do. Consequently, they need extensive evaluation.

Argumentation-based models attempt mainly to marry the exchange of offers with the exchange of arguments. This permits great flexibility since, for instance, it makes possible to persuade agents to change their view of an offer during the course of negotiation. However, most models make considerable demands on any implementation, mainly because they appeal to very rich representations of the agents and their environments. Accordingly, some researchers pointed out that agents which can argue in support of their negotiations will only ever represent a small, though important, class of automated negotiators [7].

Automated negotiation promises a higher level of process efficiency and a higher quality of agreements (when compared to traditional, face-to-face negotiation). In practice, the task of designing and implementing autonomous negotiating agents involves the consideration of insights from multiple relevant research areas to integrate different perspectives on negotiation [1]. Yet, most existing models primarily use either game-theoretic techniques or methods from the social sciences as a basis to develop negotiating agents, and largely ignore the integration of the results from both areas.

This paper argues that an interdisciplinary approach towards the development of autonomous negotiating agents is possible and highly desirable – game-theoretic (strategic) and behavioural negotiation theories can mutually reinforce each other and lead to more comprehensive and richer models. As a starting point for this research effort, this paper presents a model for autonomous agents that handles bilateral multi-issue negotiation, introduces equilibrium strategies for the bargaining game of alternating offers, and formalizes a set of strategies and tactics frequently used by human negotiators. On the one hand, it considers two fully informed agents about the various aspects of the bargaining game and employs game-theoretic techniques to define equilibrium strategies. On the other hand, it considers two incompletely informed agents about the various aspects of the game and formalizes a set of negotiation strategies and tactics studied in the social sciences.

This paper builds on our previous work in the area of negotiation [10,11,12]. In particular, it extends our negotiation model by introducing a number of negotiation strategies and tactics motivated by human procedures typical of multi-issue negotiation. It also lays the foundation for performing an experiment to investigate the performance of agents equipped with the model in terms of quality and cost of bargaining.

The remainder of the paper is structured as follows. Section 2 presents the negotiation model. In particular, subsection 2.1 describes the negotiation protocol and the preferences of the negotiators, subsection 2.2 introduces equilibrium strategies for the bilateral multi-issue bargaining game of alternating offers, and subsection 2.3 formalizes concession and problem solving strategies frequently used by human negotiators. Section 3 discusses related work and section 4 presents concluding remarks.

2   The Negotiation Model

Negotiation is usually understood as proceeding through three phases [9]: a beginning or initiation phase, a middle or problem-solving phase, and an ending or resolution phase. The initiation phase focuses on preparation and planning for negotiation and is marked by each party's efforts to posture for positions. The problem-solving phase seeks a solution for the dispute and is characterized by movement toward a final agreement. The resolution phase focuses on implementing the final agreement.

Effective preparation and planning involves the creation of a well-laid plan specifying the activities that negotiators should attend to before starting to negotiate [17,19]. That plan, and the confidence derived from it, is often a critical factor for achieving negotiation objectives. Accordingly, we describe below various activities that negotiators make efforts to perform in order to carefully prepare and plan for negotiation (see our earlier work for an in-depth discussion [10]).

Let $Ag = \{ag_1, ag_2\}$ be the set of autonomous negotiating agents. Let $Agenda = \{is_1, \ldots, is_n\}$ be the negotiating agenda — the set of (independent) issues at stake. The issues are quantitative variables, defined over continuous intervals. Effective planning requires that negotiators prioritize the issues, define the limits, and specify the targets. Priorities are set by rank-ordering the issues, *i.e.*, by defining the most important, the second most important, and so on. The priority pr_{il} of an agent $ag_i \in Ag$ for each issue $is_l \in Agenda$ is a number that represents its order of preference. The weight w_{il} of is_l is a number that represents its relative importance. The limit lim_{il} is the point where ag_i decides that it should stop to negotiate, because any settlement beyond this point is not minimally acceptable. The level of aspiration or target point trg_{il} is the point where ag_i realistically expects to achieve a settlement.

Additionally, effective planning requires that negotiators agree on an appropriate protocol that defines the rules governing the interaction. The protocol can be simple, allowing agents to exchange only proposals. Alternatively, the protocol can be sophisticated, allowing agents to provide arguments to support their negotiation stance. As noted earlier, most sophisticated protocols make considerable demands on any implementation. Thus, in this work we consider a simple protocol (see subsection 2.1).

Finally, effective planning requires that negotiators be able to select appropriate strategies that account for their individual actions. Traditionally, AI researchers have paid little attention to this pre-negotiation step. In the last several years, however, a number of researchers have developed models that include libraries of negotiation strategies (see, e.g., [6,7,8]). Some strategies are in equilibrium, meaning that no designer will benefit by building agents that use any other strategies when it is known that some agents are using equilibrium strategies (see, e.g., [14] for an in-depth description of the standard game-theoretic concept of equilibrium). Thus, for some situations of complete information, the agents can be designed to adopt equilibrium strategies (see subsection 2.2). However, for situations of incomplete information, the problem of strategic choice is rather complex. In these situations, many bargaining models have different equilibria sustained by different assumptions on what an individual in the game would believe if its opponent took an action that it was not supposed to take in equilibrium. Hence, our study differs from this line of work — we address the challenge of building agents that are able to negotiate under incomplete information by formalizing relevant strategies used by human negotiators and empirically evaluating the effectiveness of these strategies in different situations (see subsection 2.3 and the comments on future work in section 4).

2.1   The Negotiation Protocol and Time Preferences

The negotiation protocol is an alternating offers protocol [14]. Two agents or players bargain over the division of the surplus of $n \geq 2$ issues (goods or pies). The players determine an allocation of the issues by alternately proposing offers at times in $T = \{1, 2, \ldots\}$. This means that one offer is made per time period $t \in T$, with an agent, say ag_i, offering in odd periods $\{1, 3, \ldots\}$, and the other agent ag_j offering in even periods $\{2, 4, \ldots\}$. The negotiation procedure, labelled the "joint-offer procedure", involves bargaining over the allocation of the entire endowment stream at once. An offer is a vector $(x_1, \ldots, x_n)$ specifying a division of the n goods. Once an agreement is reached, the agreed-upon allocations of the goods are implemented. This procedure permits agents to exploit the benefits of trading concessions on different issues.

The players' preferences are modelled by assuming that each player ag_i discounts future payoffs at some given rate δ_i^t, $0 < \delta_i^t < 1$, (δ_i^t is referred to as the discount factor and the preferences as time preferences with a constant discount rate). The cost of bargaining derives from the delay in consumption implied by a rejection of an offer. Practically speaking, the justification for this form of preferences takes into account the fact that money today can be used to make money tomorrow. Let U_i be the payoff function of ag_i. For simplicity and tractability, we assume that U_i is separable in all their arguments and that the per-period delay costs are the same for all issues:

$$U_i(x_1, \ldots, x_n, t) = \delta_i^{(t-1)} \sum_{l=1}^{n} w_{il} \, u_{il}(x_l)$$

where w_{il} is the weight of is_{il} and x_l denotes the share of ag_i for is_{il}. The component payoff function u_{il} for is_{il} is a continuous, strictly monotonic, and linear function. The distinguish feature of time preferences with a constant discount rate is the linearity of the function u_{il} [14]. The payoff of disagreement is normalized at 0 for both players.

2.2   Equilibrium Strategies

The negotiation process is modelled as an extensive game. For theoretical convenience, we consider the standard game-theoretic situation of two players completely informed about the various aspects of the game. The players are assumed to be rational, and each player knows that the other acts rationally. Also, we consider settings in which there is more than one issue and the players have different evaluations of the issues.

First, consider a two-sided four-issue bargaining situation. Two players are jointly endowed with a single unit of each of four goods, $\{X_1, \ldots, X_4\}$, and alternate proposals until they find an agreement. Each good is modelled as an interval $[0, 1]$ (or as a divisible pie of size 1). The players' preferences are as follows:

$$U_i = \delta_i^{(t-1)} (a\,x_1 + b\,x_2 + x_3 + x_4)$$
$$U_j = \delta_j^{(t-1)} [(1 - x_1) + (1 - x_2) + c\,(1 - x_3) + d\,(1 - x_4)]$$

where x_l and $(1 - x_l)$, $l=1,\ldots,4$, denote the shares of ag_i and ag_j for each pie, respectively. The parameters a, b, c, and d allow the marginal utilities of the players to differ across issues and players. We consider $a > b > 1$ and $d > c > 1$, i.e., ag_i places greater emphasis on goods X_1 and X_2 while ag_j values goods X_3 and X_4 more. Also, we consider that δ_i and δ_j are close to 1 and the parameters a, b, c, and d are close to one another. Let $p_{j \to i}^{t-1}$ and $p_{i \to j}^{t}$ denote the offers that ag_j proposes to ag_i in period $t-1$ and ag_i proposes to ag_j in period t, respectively. Consider the following strategies:

$$str_i^* = \begin{cases} \text{offer } (1,\,1,\,x_{i3}^*,\,0) & \text{if } ag_i\text{'s turn} \\ \text{if } U_i(p_{j \to i}^{t-1}) \geq U_i^* \quad \text{accept else reject} & \text{if } ag_j\text{'s turn} \end{cases}$$

$$str_j^* = \begin{cases} \text{offer } (1,\,x_{j2}^*,\,0,\,0) & \text{if } ag_j\text{'s turn} \\ \text{if } U_j(p_{i \to j}^{t}) \geq U_j^* \quad \text{accept else reject} & \text{if } ag_i\text{'s turn} \end{cases}$$

where $U_i^* = U_i(1, x_{j2}^*, 0, 0)$, $U_j^* = U_j(1, 1, x_{i3}^*, 0)$, and the shares are the following: $x_{i3}^* = \frac{\delta_i \delta_j (a+b) - \delta_j (a+b+bc+bd) + bc + bd}{bc - \delta_i \delta_j}$ and $x_{j2}^* = \frac{\delta_i (\delta_i \delta_j (a+b) - \delta_j (a+b+bc+bd) + bc + bd) + (bc - \delta_i \delta_j)(a\delta_i + b\delta_i - a)}{b(bc - \delta_i \delta_j)}$.

Remark 1. For the two-sided four-issue bargaining game of alternating offers with an infinite horizon, in which the players' preferences are as described above, the pair of strategies (str_i^*, str_j^*) form an equilibrium. The outcome is the following:

$$x_1^* = 1, \quad x_2^* = 1, \quad x_3^* = \frac{\delta_i \delta_j (a + b) - \delta_j (a + b + bc + bd) + bc + bd}{bc - \delta_i \delta_j}, \quad x_4^* = 0$$

Agreement is immediately reached with no delay. The outcome is Pareto optimal. Letting $\delta_i \to 1$ and $\delta_j \to 1$, the equilibrium division is $(1, 1, 0, 0)$.

Now, consider a two-sided n-issue bargaining situation. Two players bargain over n distinct goods, $\{X_1, \ldots, X_n\}$, and are allowed to alternate proposals until they find an agreement. Their preferences are as defined in subsection 2.1. Again, each good is modelled as a divisible pie of size 1. The players set different weights for the goods such that: $w_{i1}/w_{j1} > w_{i2}/w_{j2} > \ldots > w_{in}/w_{jn}$.

Remark 2. The bilateral multi-issue bargaining game of alternating offers with an infinite horizon, in which the players' preferences are as described above, has an equilibrium. The outcome is Pareto optimal:

$$(x_1^*, \ldots, x_{k-1}^*, x_k^*, x_{k+1}^*, \ldots, x_n^*) = (1, \ldots, 1, s, 0, \ldots, 0)$$

where x_l^*, $l=1,\ldots,n$, denotes the share of ag_i for each divisible pie. The constant s represents the share of ag_i for the X_k pie.

The proofs of Remark 1 and Remark 2 are based on the familiar necessary conditions for equilibrium: ag_i is indifferent between waiting one period to have its offer accepted and accepting ag_j's offer immediately, and ag_j is indifferent between waiting one period to have its offer accepted and accepting ag_i's offer immediately. For instance, consider the n-issue bargaining situation. Let $\mathbf{x}_i^* = (x_{i1}^*,\ldots,x_{in}^*)$ and $\mathbf{x}_j^* = (x_{j1}^*,\ldots,x_{jn}^*)$ be the equilibrium proposals of ag_i and ag_j, respectively. The problem for ag_i is stated as follows:

$$\text{maximize:} \quad U_i(x_1,\ldots,x_n,t) = \delta_i^{(t-1)} \sum_{l=1}^{n} w_{il}\, x_l$$

$$\text{subject to:} \quad U_j(x_{i1}^*,\ldots,x_{in}^*,t) = U_j(x_{j1}^*,\ldots,x_{jn}^*,t+1)$$

$$0 \leq x_{il}^* \leq 1, \; 0 \leq x_{jl}^* \leq 1, \; \text{for} \; l=1,\ldots,n$$

This maximization problem is similar to the continuous (or fractional) knapsack problem and solvable by a greedy approach [13]. First, ag_i gives away the maximum possible share of the issue with the lowest ratio of weights. If the supply of that issue is exhausted, it gives away the maximum possible share of the issue with the next lowest ratio of weights, and so forth until ag_j gets the utility of $U_j(x_{j1}^*,\ldots,x_{jn}^*,t+1)$. The problem for ag_j is stated in a similar way and also solvable by a greedy approach.

2.3   Concession and Problem Solving Strategies

Game theory can provide sound design principles for computer scientists. The last subsection has considered two fully informed agents and used game-theoretic techniques to define equilibrium strategies. The agents were creative, honest, and able to settle for the outcome that maximizes their benefit (resources were not wasted and money was not squandered). Yet, the assumption of complete information is of limited use to the designers of agents. In practice, agents have private information. Also, simple casual observation reveals the existence of concessions and long periods of disagreement in many actual negotiations. Furthermore, one agent, say ag_i, may wish to act rationally, but the other agent may not behave as a strategically sophisticated, utility maximizer — thus rendering conventional equilibrium analysis inapplicable.

Behavioral negotiation theory can provide rules-of-thumb to agent designers. The danger is that the designers may not be fully aware of the circumstances to which human practice is adapted, and hence use rules that can be badly exploited by new agents. Nevertheless, an increasing number of researchers consider that human practice is crucial to automated negotiation (see, e.g., [1]). There is a need to integrate the procedures and results from behavioral negotiation theory in bargaining models incorporating game-theoretic techniques. Accordingly, this subsection considers two incompletely informed agents about the various aspects of the bargaining game and formalizes relevant strategies studied in the social sciences.

Negotiation strategies can reflect a variety of behaviours and lead to strikingly different outcomes. However, the following two fundamental groups of strategies are commonly discussed in the behavioral negotiation literature [15,19]:

1. *concession making* — negotiators who employ strategies in this group reduce their aspirations to accommodate the opponent;
2. *problem solving* — negotiators maintain their aspirations and try to find ways of reconciling them with the aspirations of the opponent.

Although it is important to distinguish among these two groups of strategies, we hasten to add several explanatory notes. First, most strategies are implemented through a variety of tactics. The line between strategies and tactics often seems indistinct, but one major difference is that of scope. Tactics are short-term moves designed to enact or pursue broad (high-level) strategies [9]. Second, concession making strategies are essentially unilateral strategies — the decision to concede is fundamentally a unilateral one. By contrast, problem solving strategies are essentially social strategies. Third, most negotiation situations call forth a combination of strategies from different groups. Finally, most strategies are only informally discussed in the behavioral literature. They are not formalized, as typically happens in the game-theoretic literature.

Concession making behaviour aims at partially or totally accommodating the other party. Consider two incompletely informed agents bargaining over n distinct issues $\{is_1, \ldots, is_n\}$. For convenience, each issue is_l is modelled as an interval $[min_l, max_l]$. The agents' preferences are as defined in subsection 2.1. The opening stance and the pattern of concessions are two central elements of negotiation. Three different opening positions (extreme, reasonable and modest) and three levels of concession magnitude (large, moderate and small) are commonly discussed in the behavioral literature [9]. They can lead to a number of concession strategies, notably:

1. *starting high and conceding slowly* — negotiators adopt an optimistic opening attitude and make successive small concessions;
2. *starting reasonable and conceding moderately* — negotiators adopt a realistic opening attitude and make successive moderate concessions.

Let $p_{j \to i}^{t-1}$ be the offer that ag_j has proposed to ag_i in period $t-1$. Likewise, let $p_{i \to j}^t$ be the offer that ag_i is ready to propose in the next time period t. The formal definition of a generic concession strategy follows.

Definition 1. *Let $ag_i \in Ag$ be a negotiating agent. A concession strategy for ag_i is a function that specifies either the tactic to apply at the beginning of negotiation or the tactic that defines the concessions to be made during the course of negotiation:*

$$
conc \stackrel{def}{=} \begin{cases}
apply\ tact_i^1 & \text{if } ag_i\text{'s turn and } t=1 \\
apply\ tact_i^t & \text{if } ag_i\text{'s turn and } t>1 \\
if\ U_i(p_{j \to i}^{t-1}) \geq U_i(p_{i \to j}^t) \quad accept\ else\ reject & \text{if } ag_j\text{'s turn}
\end{cases}
$$

where $tact_i^1$ is an opening negotiation tactic and $tact_i^t$ is a concession tactic.

The two aforementioned concession strategies are defined by considering different tactics. For instance, the "starting reasonable and conceding moderately" strategy is defined by: "$tact_i^1 = starting\_realistic$" and "$tact_i^t = moderate$" (but see below).

Problem solving behaviour aims at finding agreements that appeal to all sides, both individually and collectively. The host of problem solving strategies includes [15]:

1. *low-priority concession making* — negotiators hold firm on more important issues while conceding on less important issues;
2. *logrolling* — negotiators agree to trade-off among the issues under consideration so that each party concedes on issues that are of low priority to itself and high priority to the other party.

Low-priority concession making involves primarily the analysis of one's priorities and further concessions on less important issues. However, effective logrolling requires information about the two parties' priorities so that concessions can be matched up. This information is not always easy to get. The main reason for this is that negotiators often try to conceal their priorities for fear that they will be forced to concede on issues of lesser importance to themselves without receiving any repayment [15]. Despite this, research evidence indicates that it is often not detrimental for negotiators to disclose information that can reveal their priorities — a simple rank order of the issues does not put negotiators at a strategic disadvantage [19]. Hence, we consider that negotiators willingly disclose information that can help to identify their priorities (e.g., their interests).

Logrolling can be viewed as a variant of low-priority concession making in which the parties' priorities are in the opposite direction. The formal definition of a generic logrolling strategy follows (the definition of a low-priority concession making strategy is essentially identical, and is omitted).

Definition 2. *Let $ag_i \in Ag$ be a negotiating agent and $ag_j \in Ag$ be its opponent. Let Agenda denote the negotiating agenda, $Agenda^{\oplus}$ the subset of the agenda containing the issues of high priority for ag_i (and low priority for ag_j), and $Agenda^{\ominus}$ the subset of the agenda containing the issues of low priority for ag_i (and high priority for ag_j). A logrolling strategy for ag_i is a function that specifies either the tactic to apply at the beginning of negotiation or the tactics to make trade-offs during the course of negotiation:*

$$log \overset{def}{=} \begin{cases} apply\ tact_i^1 & if\ ag_i\text{'s turn and } t=1 \\ apply\ tact_i^{t\oplus}\ and\ tact_i^{t\ominus} & if\ ag_i\text{'s turn and } t>1 \\ if\ U_i(p_{j\to i}^{t-1}) \geq U_i(p_{i\to j}^t)\quad accept\ else\ reject & if\ ag_j\text{'s turn} \end{cases}$$

where $tact_i^1$ is an opening negotiation tactic, $tact_i^{t\oplus}$ is a concession tactic (to apply to the issues on $Agenda^{\oplus}$), and $tact_i^{t\ominus}$ is another concession tactic (to apply to the issues on $Agenda^{\ominus}$). ■

A number of logrolling strategies can be defined simply by considering different tactics. For instance, a strategy that specifies a realistic opening attitude, followed by null concessions on issues on $Agenda^{\oplus}$, and large concessions on issues on $Agenda^{\ominus}$, is defined by: "$tact_i^1 = starting\_realistic$", "$tact_i^{t\oplus} = stalemate$", and "$tact_i^{t\ominus} = soft$" (but see below).

Opening negotiation tactics are functions that specify the initial values for each issue is_l at stake. The following three tactics are commonly discussed in the behavioral literature [9]:

1. *starting optimistic* — specifies a value far from the target point;
2. *starting realistic* — specifies a value close to the target point;
3. *starting pessimistic* — specifies a value close to the limit.

The definition of the tactic "starting realistic" follows (the definition of the other two tactics is essentially identical, and is omitted).

Definition 3. *Let $ag_i \in Ag$ be a negotiating agent and $is_l \in Agenda$ a negotiation issue. Let trg_{il} be the target point of ag_i for is_l. The tactic starting realistic for ag_i is a function that takes is_l and trg_{il} as input and returns the initial value $v[is_l]_i^1$ of is_l:*

$$starting\_realistic(is_l, trg_{il}) = v[is_l]_i^1$$

where $v[is_l]_i^1 \in [trg_{il} - \epsilon, \; trg_{il} + \epsilon]$ and $\epsilon > 0$ is small. ∎

Concession tactics are functions that compute new values for each issue is_l. The following five tactics are commonly discussed in the literature [9]:

1. *stalemate* — models a null concession on is_l;
2. *tough* — models a small concession on is_l;
3. *moderate* — models a moderate concession on is_l;
4. *soft* — models a large concession on is_l;
5. *accommodate* — models a complete concession on is_l.

The definition of a generic concession tactic follows (without loss of generality, we consider that ag_i wants to maximize is_l).

Definition 4. *Let $ag_i \in Ag$ be a negotiating agent, $is_l \in Agenda$ a negotiation issue, and lim_{il} the limit of is_l. Let $v[is_l]_i^t$ be the value of is_l offered by ag_i at period t. A concession tactic for ag_i is a function that takes $v[is_l]_i^t$, lim_{il} and the concession factor $Cf \in [0, 1]$ as input and returns the new value $v[is_l]_i^{t+2}$ of is_l:*

$$concession\_tactic(v[is_l]_i^t, lim_{il}, Cf) = v[is_l]_i^{t+2}$$

where $v[is_l]_i^{t+2} = v[is_l]_i^t - Cf\,(v[is_l]_i^t - lim_{il})$. ∎

The five tactics are defined by considering different values for Cf. In particular, the stalemate tactic by $Cf = 0$, the accommodate tactic by $Cf = 1$, and the other three tactics by different ranges of values for Cf (e.g., the tough tactic by $Cf \in\,]0.00, 0.05]$, the moderate tactic by $Cf \in\,]0.05, 0.10]$, and the soft tactic by $Cf \in\,]0.10, 0.15]$).

3   Related Work

AI researchers have investigated the design of autonomous negotiating agents from two main perspectives: a theoretical or formal perspective and a practical or computational perspective. Researchers following the theoretical perspective attempt mainly to develop formal models for describing, specifying, and reasoning about the key features of the agents. Most researchers have focused on formal bargaining, auctions, market-oriented programming, contracting, and coalition formation (see, e.g., [5,8]). On the other hand, researchers following the practical perspective attempt mainly to develop computational models for specifying the key data structures of the agents and the processes operating on these structures. Most models are based on informal procedures from the social sciences (see, e.g., [3,6,10]).

The task of designing and implementing negotiating agents involves the consideration of insights from multiple relevant research areas to integrate different perspectives on negotiation. In particular, game-theoretic (strategic) and behavioural negotiation theories can mutually reinforce each other and lead to more comprehensive and richer models. Yet, the majority of existing models largely ignore the integration of the results from both research areas. Noting this gap, this paper uses both game-theoretic techniques and methods from the social sciences as a basis to develop negotiating agents.

4 Conclusion

In this paper, we argue that an interdisciplinary approach towards the development of autonomous negotiating agents is possible and highly desirable. As a starting point for this research effort, we present a model that handles bilateral multi-issue negotiation, employs game-theoretic techniques to define equilibrium strategies for the bargaining game of alternating offers, and formalizes a set of negotiation strategies studied in the social sciences. Autonomous agents equipped with the model are currently being developed using the Jade framework. The agents are able to negotiate under both complete and incomplete information, thereby making the model in particular and the agents in general very compelling for automated negotiation.

Our aim for the future is to extend the model and to perform its experimental validation. In particular, the model defines a number of strategies based on rules-of-thumb distilled from behavioral negotiation theory. Hence, these strategies need to be empirically evaluated to determine precisely how they behave in different situations. In addition, we intend to study the bargaining game of alternating offers in order to define equilibrium strategies for two incompletely informed players.

References

1. Bichler, M., Kersten, G., Strecker, S.: Towards a Structured Design of Electronic Negotiations. Group Decision and Negotiation 12, 311–335 (2003)
2. Bussmann, S., Jennings, N., Wooldridge, M.: Multiagent Systems for Manufacturing Control. Springer, Heidelberg (2004)
3. Faratin, P.: Automated Service Negotiation Between Autonomous Computational Agents, Ph.D. Thesis, Queen Mary & Westfield College, UK (2000)
4. Fasli, M.: Agent Technology for e-Commerce. John Wiley & Sons, Chichester (2007)
5. Fatima, S., Wooldridge, M., Jennings, N.: Multi-Issue Negotiation with Deadlines. Journal of Artificial Intelligence Research 27, 381–417 (2006)
6. Ito, T., Hattori, H., Zhang, M., Matsuo, T.: Rational, Robust, and Secure Negotiations in Multi-Agent Systems. Springer, Heidelberg (2008)
7. Jennings, N., Faratin, P., Lomuscio, A., Parsons, S., Wooldridge, M., Sierra, C.: Automated Negotiation: Prospects, Methods and Challenges. Group Dec. and Neg. 10, 199–215 (2001)
8. Kraus, S.: Strategic Negotiation in Multi-Agent Environments. MIT Press, Cambridge (2001)

9. Lewicki, R., Barry, B., Saunders, D., Minton, J.: Negotiation. McGraw Hill, New York (2003)
10. Lopes, F., Mamede, N., Novais, A.Q., Coelho, H.: A Negotiation Model for Autonomous Computational Agents: Formal Description and Empirical Evaluation. Journal of Intelligent & Fuzzy Systems 12, 195–212 (2002)
11. Lopes, F., Mamede, N., Novais, A.Q., Coelho, H.: Negotiation Strategies for Autonomous Computational Agents. In: ECAI 2004, pp. 38–42. IOS Press, Amsterdam (2004)
12. Lopes, F., Novais, A.Q., Coelho, H.: Interdisciplinary Approach to Automated Negotiation: A Preliminary Report. In: ECMS 2008. SCS Publishing House (to appear, 2008)
13. Martello, S., Toth, P.: Knapsack Problems: Algorithms and Computer Implementations. John Wiley & Sons, Chichester (1990)
14. Osborne, M., Rubinstein, A.: Bargaining and Markets. Academic Press, London (1990)
15. Pruitt, D., Kim, S.: Social Conflict: Escalation, Stalemate, Settlem. McGraw Hill, New York (2004)
16. Rahwan, I., Ramchurn, S., Jennings, N., McBurney, P., Parsons, S., Sonenberg, L.: Argumentation-based Negotiation. The Knowledge Engineering Review 18, 343–375 (2004)
17. Raiffa, H.: The Art and Science of Negotiation. Harvard University Press (1982)
18. Simon, H.: The Sciences of the Artificial. MIT Press, Cambridge (1981)
19. Thompson, L.: The Mind and Heart of the Negotiator. Prentice-Hall, Englewood Cliffs (2005)

Bargaining Power in Electronic Negotiations:
A Bilateral Negotiation Mechanism

Ricardo Buettner[1] and Stefan Kirn[2]

[1] Fachhochschule fuer Oekonomie & Management - University of Applied Sciences,
80636 Muenchen, Germany
`ricardo.buettner@fom.de`
[2] Information Systems II, University of Hohenheim, 70593 Stuttgart, Germany
`stefan.kirn@uni-hohenheim.de`

Abstract. Bargaining power has a major influence in negotiations. Up to now, a lot of electronic negotiation models have been developed and manifold negotiation challenges have been already addressed, but mainly related to the structure and the process of the negotiation. However, research concerning bargaining power is still inadequate represented. Thus, in order to contribute to the state of the art of electronic negotiations, this paper shows a bilateral automated negotiation mechanism that considers bargaining power.

1 Motivation and Problem Description

In spite of a rich developed field of electronic negotiation models, the best initial negotiation offer is still a major challenge. The corresponding problem is the following: Generally, whenever an (initial) offer is made by a negotiation partner, it needs a good reason to vary it. So he is more or less bounded to his offer. If the initial offer is too far away from a potential zone of agreement in order to make a good deal, it will normally take a long time to reach an agreement and some of the negotiation partners will probably exit the negotiation before. However, if the initial offer is too close to the reservation price of the bidder at a risen possibility of a quick agreement, the surplus of the bidder will be smaller. The best initial negotiation offer depends on many factors; for example negotiation deadlines [41], utility functions [17] etc. However, research concerning bargaining power, as one of this major influencing factors [13,36], is still inadequate represented [11,19]. Thus, this paper focuses on bargaining power in electronic negotiations.

The text of the paper is divided into 4 parts: In section 2 we sketch the role of bargaining power in organizational approaches and related work in multi-agent literature. Section 3 presents an overview of electronic negotiations. After that, section 4 shows a bilateral automated negotiation mechanism considering bargaining power. At the end, future research questions are introduced.

2 Power in Organizational Approaches

A lot of old and modern scientific theories point out the significant influence of the power distribution in organizations (e. g., Contingency-Theory, Behavioral-Approach).

G. Psaila and R. Wagner (Eds.): EC-Web 2008, LNCS 5183, pp. 92–101, 2008.

2.1   Contingency-Theory

The Contingency approach is rooted in different other theories. For example, research in social sciences pointed out within the scope of the Bureaucracy Approach by *M. Weber* [56] that organizational structures are not in line with the ideal type of bureaucracy. During the fifties business studies came to the conclusion that a universal organization structure can not exist. One of the first was *J. Woodward* [57,58], who argued the importance of taking the individual initial position of the organization into account whenever design recommendations concerning the organizational structure are given. At the beginning of the development of the Contingency-Theory, empirical research studied the impact of the influencing factors to the organization structure [12]. For example, following the popular thesis by *T. Burns* and *G. M. Stalker*, organic structures are in line with dynamic environments and mechanistic structures fit to static ones [8].

The Contingency-Theory presumes power in organizations as an important phenomenon. The management tries to make sure its authority, mainly on the basis of organizational rules.

2.2   Behavioral Approach

The development of the behavioral approach is marked by the Bayes-Rule [2], the "St. Petersburg-Game" [4] by *D. Bernoulli*, the Expected Utility Theory by *J. L. von Neumann* and *O. Morgenstern* [55], the concept of Bounded Rationality by *H. A. Simon* [49], the concept of Satisficing [35,49], the Subjective Expected Utility Theory by *L. J. Savage* [45], the Allais-Paradox [1], the Ellsberg-Paradox [16], the Prospect-Theory by *D. Kahneman* and *A. Tversky* [24,25], the Regret-Theory by *G. Loomes* and *R. Sugden* [33] or rather by *D. E. Bell* [3] and the concept of Framing by *A. Tversky* and *D. Kahneman* [53,54].

The behavioral approach shows key findings concerning the rationality assumptions in negotiations. Recapitulating, the assumption of perfect rationality (Homo Oeconomicus) was theoretically and empirically disproved. The neoclassic premises were revised by the behavioral approach and replaced by bounded rationality and imperfect information situations [49]. Therewith, bargaining power plays an important role within negotiations [41]. The decision process is significantly influenced by the distribution of power.

2.3   Power in Multi-Agent-Systems

M. S. Fox [18], *T. W. Malone* [34], *K. M. Carley* and *L. Gasser* [10], *et al.* consider Multi-Agent-Systems (MAS) as organizations (in the sense of organized social entities). Availability of different forms of power (e. g., resources, information, coalitions [27]) have been identified since the eighties [42]. It is also well understood that intelligent agents are driven by intentions (e. g., BDI [7]) and that they plan and perform actions in order to meet their objectives. To this purpose, they aim to make optimal use of their resources, which also includes different forms of power. So far, however, there is only little work on integration of power distribution into negotiation protocols.

3   A Brief Description of Electronic Negotiations

Electronic automating of negotiations was forecasted by [14] more than 20 years ago. However, the <u>automation level</u> of present negotiation systems is still different: Full-automated, process support and hybrid negotiation models exist today. Full-automated models work without any human interaction and are strictly structured. Process support models (e. g., INSPIRE [26] or www.ebay.com/) only facilitate the negotiation. Hybrid models are partly-automated, for example [15]. The literature shows many definitions for (electronic) negotiations; e. g. [32,43]. A pragmatic way to define a negotiation is the following [5, p. 316]:

Definition 1. *A negotiation is an iterative communication and decision making process between two or more participants who: (1) cannot achieve their objectives through unilateral actions, (2) exchange information comprising offers, counter-offers and arguments, (3) deal with interdependent tasks, and (4) search for a consensus which is a compromise decision.*

J. S. Rosenschein and *G. Zlotkin* were the first who analyzed strategic interactions between self-interested agents [42,43,60] on the basis of the fundamental game-theoretic work [55] by *J. L. von Neumann* and *O. Morgenstern*. Their formal analysis is based on the distributed problem solving approach (see [6]) adapted from the work of *J. C. Harsanyi et al.* [20,21] and *D. M. Kreps et al.* [30]. The underlying <u>game-theoretic approach</u> studies the equilibrium conditions and tries to find out the optimal strategy between identical agents [31,37,38,39,46,47,48]. Game-theoretic models are deemed to be mathematically elegant, but they are very restricted in use because of their assumptions of perfect rationality, unlimited resources and a perfect information situation [22,23,40]. In order to relax this restrictions, heuristic approaches have been adapted for electronic negotiations. <u>Heuristic approaches</u> solve the problematic assumption of unlimited resources by using thumb rules (e. g., [28]). Thus, the assumption of perfect rationality is also rejected. But, electronic negotiation models on the basis of heuristic approaches need an intensive evaluation, regular via simulation or empirical investigations [22, p. 210]. Finally <u>argumentation-based negotiations</u> have been developed. There, the agents have the possibility to reason about their positions. When the negotiation partner is persuaded, it will change its negotiation position. The argumentation-based approach increases the possibility and the quality of an agreement compared to game-theoretic or heuristic solutions and was firstly realized by *K. P. Sycara* [52] (negotiation support system PERSUADER).

Electronic negotiation models can handle many organizational challenges [9]. For example, regarding the negotiation structure, bilateral, one-sided multilateral and double-sided multilateral negotiations are separated (<u>protocol category</u>) [5]. Bilateral negotiations are restricted to two negotiation partners (one buyer and one seller) and were first analyzed by [14,50]. One-sided multilateral negotiations are deemed to be the standard form of auctions and are either characterized by one seller and many buyers or vice versa. Finally, double-sided multilateral negotiations are characterized by many buyers and many sellers (e. g., [59]). Depending on how many attributes for the negotiation item are taken into account during the evaluation, two attribute types are distinguished: In the single-attribute case the negotiation item is evaluated only by one characteristic, normally the price. All other attributes, for instance quality or warranty, are agreed in advance and will not be negotiated. In contrast, in multi-attribute negotiations more than one characteristic are taken simultaneously into account, e. g., [26]. Beyond, the number of positions describes the quantity of independent items in a single negotiation over there a final decision is made. Electronic negotiation systems with the possibility to contract a high number of positions (e. g., [15]) are of significant practical relevance. Furthermore, negotiations can be separated in non-mediated (e. g., INSPIRE [26]) and mediated negotiations.

In addition to the negotiation structure a lot of process-related challenges have been already addressed. Firstly, a negotiation can be separated into public and closed sessions [5, p. 318]. As in public negotiations new participants can take part dynamically, this is not allowed in closed sessions. Furthermore, electronic negotiations can be distinguished by the binding type. Binding negotiations (e. g., www.ebay.com/) ask for an authentication of every participant in advance. Finally, time has an enormous influence to negotiations [41,51]: At first, S. Kraus et al. [29] took time limits into account in electronic negotiations.

4   A Negotiation Mechanism for Asymmetric Bargaining Power Distribution

As shown in the last section, many negotiation challenges have been already addressed. However, works studying behavioral issues, especially concerning bargaining power, are still inadequate represented. Thus, this section presents a negotiation mechanism focussing on bargaining power.

4.1   Initial Assumptions

1. There are two agents, A and B. A is the seller and B the buyer agent.
2. Each agent wants to maximize its own utility ($utility_A$ and $utility_B$).
3. The distribution of the bargaining power between A and B is different ($power_A$ and $power_B$).
4. Each agent knows all relevant information, except the reservation price of the other.

4.2   Protocol Design

The protocol is based on the general model of alternating offers (Fig. 1) [44]. At time t, one of the agents A or B can make one of the following moves: (a) make a new offer or (b) accept or decline the offer of the negotiation partner or (c) exit the negotiation. The reservation price is the price below (above) which a seller (buyer) is unwilling to go. It is private and different for each agent. As shown in Figure 1, when the reservation price of the seller (RP_B)

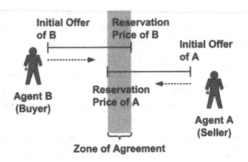

Fig. 1. Bilateral model of alternating offers

is lower than the reservation price of the buyer (RP_A), any point within the "zone of agreement" is an appropriate solution. If $RP_B > RP_A$, the zone of agreement does not exist and no deal can be reached via negotiation. If a zone of agreement exists, typically both the buyer and the seller will make concessions from their initial proposal. The buyer will increase its initial proposal, while the seller will decrease it. Eventually, a proposal within the zone of agreement will be acceptable to both. The utility of the agents are defined as following:

Definition 2. *The utility$_A$ of the seller agent is the difference between the equilibrium price P and its reservation price RP_A (utility$_A = P - RP_A$).*

Definition 3. *The utility$_B$ of the buyer agent is the difference between its reservation price RP_B and the equilibrium price P (utility$_B = RP_B - P$).*

Assuming perfect rationality, the distribution of bargaining power is implemented as follows:

Definition 4. *The bargaining power$_{A,B}$ of agents A and B describes their respective strengths in a negotiation ($0 < power_{A,B} < 1, power_A + power_B = 1$). power$_{A,B}$ directly determine the utility distribution between the agents in equilibrium state ($\frac{power_A}{power_B} = \frac{utility_B}{utility_A}$).*

4.3 Equilibrium

It is assumed that the current offer of the seller agent A is not higher than its last offer ($bid_A^{t+1} \leq bid_A^t$). Vice versa for B: $bid_B^{t+1} \geq bid_B^t$. I.e., an agent can not make the negotiation partner an inferior offer compared to its last offer. Based on this as well as on the utility and power conditions (see Defn. 2-4), after $t \to \infty$ negotiation steps, every negotiation will end in an equilibrium state (Defn. 5) as far as there is no break-off.

Definition 5. *The equilibrium state is marked by the equilibrium price $P^{t \to \infty}$:*

$$P^{t \to \infty} = \frac{RP_A + RP_B \frac{power_A}{power_B}}{1 + \frac{power_A}{power_B}} \qquad (1)$$

4.4 Bidding Strategy

Although the negotiation agent (A or B) knows its own reservation price RP, it does not know the precise value of the other one. Accordingly, the zone of agreement is not known by either of the agents (Fig. 1). Nevertheless, the agent can observe negotiation results of comparable negotiations; especially the negotiation price p_i of a negotiation i. Therewith it knows the average negotiation price $\tilde{P} = \frac{\sum_{i=1}^{m} p_i}{m}$ ($\tilde{P}(m \rightarrow \infty) = P^{t \rightarrow \infty}$). The agent can update $\tilde{P}$ based on its own negotiation results and on its domain knowledge. As a result of this learning, the agent is expected to gain more accurate expectation of the payoff structure of the negotiation agent and therefore can make more advantageous offers. When the initial offer ($bid^{t=0}$) of an agent is close to the reservation price of the negotiation partner, the negotiation will be successfully finished soon.

From (1) follows:
$$RP_A = P + \frac{power_A}{power_B}(P - RP_B) \qquad (2)$$

Based on the observed price average ($\tilde{P}$) of comparable[1] negotiation results, the first bid ($bid_B^{t=0}$) of the buyer agent B should be the expected rounded reservation price ($\lceil R\hat{P}_A \rceil$) of A (Eqn. 3), vice verse for B (Eqn. 4):

From (2) follows:
$$bid_B^{t=0} = \lceil R\hat{P}_A \rceil = \lceil \tilde{P} + \frac{power_A}{power_B}(\tilde{P} - RP_B) \rceil \qquad (3)$$

From (2) follows:
$$bid_A^{t=0} = \lfloor R\hat{P}_B \rfloor = \lfloor \tilde{P} - \frac{power_B}{power_A}(RP_A - \tilde{P}) \rfloor \qquad (4)$$

4.5 Experimental Results and Discussion

For experimental validation the simulation setting in Table 1 was used.

Table 1. Setting for experimental results

	Agent A (Seller)	Agent B (Buyer)
Reservation Price ($RP_{A,B}$)	$RP_A = 100€$	$RP_B = 300€$
Bargaining Power ($power_{A,B}$)	$power_A = 0.2$	$power_B = 0.8$
Observed Price Average ($\tilde{P}$)	$\tilde{P} = 140€$	
Negotiation Rounds (n)	$n = 15$	

The experimental results presented in Figure 2 show that the negotiation based on the bidding strategy presented in this paper will reach a negotiated price of $bid^{t=15} = 114.72€$, close to the reservation price of the seller ($RP_A = 100€$). From the welfare perspective, the bidding strategy presented in this paper is much better compared to (initial) bids based on other strategies (Table 2).

---

[1] Comparable means in this context similar the same negotiation conditions (e.g., number of participants, distribution of negotiation power, negotiated good).

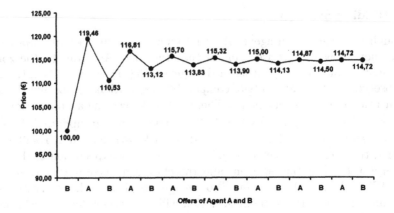

Fig. 2. Simulation results

Table 2. Experimental results

Bidding Strategy	Negotiated Price $(bid^{t=15})$	Producer Surplus (S_A)	Consumer Surplus (S_B)
Equipartition ($\tilde{P}$)	200.00€	100.00€	100.00€
Observed Price Average ($bid_B^{t=0} = \tilde{P}$)	158.24€	58.24€	141.76€
Our ($bid_B^{t=0} = RP_A$)	114.72€	14.72€	185.28€

The results explicitly show the advantages of the presented strategy. The surplus increases notedly (Fig. 3). In our experimental case, agent B (buyer) performed strongly because of its first appropriate bid based on our strategy.

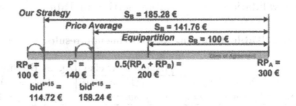

Fig. 3. Benefits of our strategy from the welfare perspective

5   Conclusion and Future Research

Despite of the fact, that many negotiation challenges have been already addressed, this paper identified the insufficient representation of bargaining power. As shown, bargaining power plays an important role in Sciences (e. g., Bureaucracy Approach, Contingency-Theory, Behavioral-Approach, Jurisprudence and Political Sciences). Thus, this paper presented a bilateral automated negotiation mechanism considering bargaining power in order to contribute to the state of the art of electronic negotiations. However, there is still a need of research. The presented mechanism is only bilateral and the utility functions are linear. Future research should relaxe this restrictions in order to show a more general model considering bargaining power. Finally, an intensive evaluation is still needed.

References

1. Allais, M.: Le Comportement de l'Homme Rationnel devant le Risque: Critique des postulats et axiomes de l'École américaine. Economtrica 21(4), 503–546 (1953)
2. Bayes, T.: An Essay Towards Solving a Problem in the Doctrine of Chances. In: Philosophical Transactions of the Royal Society of London, vol. 53, pp. 370–418 (1763)
3. Bell, D.E.: Regret in Decision Making under Uncertainty. Operations Research 30(5), 961–981 (1982)
4. Bernoulli, D.: Specimen Theoriae Novae de Mensara Sortis, pp. 175–192; translated by Sommer, L.: Exposition on a New Theory on the Measurement of Risk. Econometrica 22(1), 23–36 (1954)
5. Bichler, M., Kersten, G.E., Strecker, S.: Towards a Structured Design of Electronic Negotiations. Group Decision and Negotiation 12(4), 311–335 (2003)
6. Bond, A.H., Gasser, L.: An Analysis of Problems and Research in DAI. In: Bond, A.H., Gasser, L. (eds.) Readings in Distributed Artificial Intelligence, pp. 3–35. Morgan Kaufmann Publishers, San Mateo (1988)
7. Bratman, M.E.: Intentions, Plan and Practical Reason. Harvard University Press, Cambridge (1987)
8. Burns, T., Stalker, G.M.: The Management of Innovation. Tavistock Publ., London (1961)
9. Büttner, R.: A Classification Structure for Automated Negotiations. In: WI-IAT 2006 Workshops Proceedings, Hong Kong, China, pp. 523–530. IEEE Computer Society Press, Los Alamitos (2006)
10. Carley, K.M., Gasser, L.: Computational Organization Theory. In: Multiagent Systems: A Modern Approach to Distributed Artificial Intelligence, ch. 7, pp. 299–330. MIT Press, Cambridge (1999)
11. Castelfranchi, C.: Social Power: A Point Missed in Multi-Agent, DAI and HCI. In: Decentralized Artificial Intelligence: Proc. of the 1st European Workshop on Modelling Autonomous Agents in a Multi-Agent World (MAAMAW 1989), Cambridge, England, pp. 49–62. Elsevier, Amsterdam (1990)
12. Child, J.: More myths of management organization? Journal of Management Studies 7(3), 376–390 (1970)
13. Chung, T.-Y.: On Strategic Commitment: Contracting versus Investment. American Economic Review 85(2), 437–441 (1995)
14. Davis, R., Smith, R.G.: Negotiation as a Metaphor for Distributed Problem Solving. Artificial Intelligence 20(1), 63–109 (1983)
15. de Paula, G.E., Ramos, F.S., Ramalho, G.L.: Bilateral Negotiation Model for Agent Mediated Electronic Commerce. In: Dignum, F.P.M., Cortés, U. (eds.) AMEC 2000. LNCS (LNAI), vol. 2003, pp. 1–14. Springer, Heidelberg (2001)
16. Ellsberg, D.: Risk, ambiguity and Savage axioms. Quarterly Journal of Economics 75(4), 643–669 (1961)
17. Faratin, P., Sierra, C., Jennings, N.R.: Negotiation Decision Functions for Autonomous Agents. Robotics and Autonomous Systems 24(3-4), 159–182 (1998)

18. Fox, M.S.: An Organizational View of Distributed Systems. IEEE Transactions on Systems, Man and Cybernetics 11(1), 70–80 (1981)
19. Guyot, P., Drogoul, A., Honiden, S.: Power and Negotiation: Lessons from Agent-Based Participatory Simulations. In: Proc. of the 5th Int. Joint Conference on Autonomous Agents and Multiagent Systems, pp. 27–33. ACM Press, New York (2006)
20. Harsanyi, J.C.: Games with Incomplete Information Played by 'Bayesian' Players, I: The Basic Model. Management Science 14(3), 159–182 (1967)
21. Harsanyi, J.C., Selten, R.: A generalized Nash solution for two-person bargaining games with incomplete information. Management Science 18(5), 80–106 (1972)
22. Jennings, N.R., Faratin, P., Lomuscio, A.R., Parsons, S., Sierra, C., Wooldridge, M.: Automated Negotiation: Prospects, Methods and Challenges. Group Decision and Negotiation 10(2), 199–215 (2001)
23. Jennings, N.R., Sycara, K.P., Wooldridge, M.: A Roadmap of Agent Research and Development. Autonomous Agents and Multi-Agent Systems 1(1), 7–38 (1998)
24. Kahneman, D., Tversky, A.: Prospect Theory: An Analysis of Decision under Risk. Econometrica 47(2), 263–291 (1979)
25. Kahneman, D., Tversky, A.: Advances in Prospect Theory: Cumulative Representation of Uncertainty. Journal of Risk and Uncertainty 5(4), 297–323 (1992)
26. Kersten, G.E., Noronha, S.J.: Supporting International Negotiations with a WWW-based System, Interim Report 97-49, International Institute for Applied Systems Analysis (IIASA), Austria (August 1997)
27. Klusch, M., Shehory, O.: Coalition Formation Among Rational Information Agents. In: Perram, J., Van de Velde, W. (eds.) MAAMAW 1996. LNCS, vol. 1038, pp. 204–217. Springer, Heidelberg (1996)
28. Kraus, S.: Strategic Negotiation in Multi-Agent Environments. MIT Press, Cambridge (2001)
29. Kraus, S., Wilkenfeld, J., Zlotkin, G.: Multiagent negotiation under time constraints. Artificial Intelligence 75(2), 297–345 (1995)
30. Kreps, D.M., Milgrom, P.R., Roberts, J., Wilson, R.B.: Rational Cooperation in the Finitely Repeated Prisoners' Dilemma. Journal of Economic Theory 27(2), 245–252 (1982)
31. Kreps, D.M., Wilson, R.B.: Sequential equilibria. Econometrica 50(4), 863–894 (1982)
32. Lee, L.C.-H.: Negotiation Strategies and their Effect in a Model of Multi-Agent Negotiation. PhD thesis, Dep. of Computer Science, University of Essex (1996)
33. Loomes, G., Sugden, R.: Regret Theory: An Alternative Theory of Rational Choice under Uncertainty. Economic Journal 92(368), 805–824 (1982)
34. Malone, T.W.: Modeling Coordination in Organizations and Markets. Management Science 33(10), 1317–1332 (1987)
35. March, J.G., Simon, H.A.: Organizations. John Wiley, Chichester (1958)
36. Marwell, G., Ratcliff, K., Schmitt, D.R.: Minimizing differences in a maximizing difference game. Journal of Personality and Social Psychology 12(2), 158–163 (1969)
37. Nash, J.F.: The Bargaining Problem. Econometrica 18(2), 155–162 (1950)
38. Nash, J.F.: Equilibrium Points in n-Person Games. In: Proceedings of the National Academy of Sciences, volume 36, pp. 48–49, USA (January 1950)
39. Nash, J.F.: Non-Cooperative Games. Annals of Mathematics 54(2), 286–295 (1951)
40. Osborne, M.J., Rubinstein, A.: A Course in Game Theory. MIT Press, Cambridge (1994)

41. Pruitt, D.G., Drews, J.L.: The effect of time pressure, time elapsed, and the opponent's concession rate on behavior in negotiation. Journal of Experimental Social Psychology 5(1), 43–60 (1969)
42. Rosenschein, J.S.: Rational Interaction: Cooperation Among Intelligent Agents. PhD thesis, Computer Science Department, Stanford University, Stanford, California, USA (March 1985)
43. Rosenschein, J.S., Zlotkin, G.: Rules of Encounter: Designing Conventions for Automated Negotiation among Computers. MIT Press, Boston (1994)
44. Rubinstein, A.: Perfect Equilibrium in a Bargaining Model. Econometrica 50(1), 97–109 (1982)
45. Savage, L.J.: The Foundations of Statistics. John Wiley, New York (1954)
46. Schelling, T.C.: The Strategy of Conflict. Harvard University Press, Cambridge (1960)
47. Selten, R.: Spieltheoretische Behandlung eines Oligopolmodells mit Nachfrageträgheit. Zeitschr. für die ges. Staatswissenschaft 121, 301–324 (1965)
48. Selten, R.: Reexamination of the Perfectness Concept for Equilibrium Points in Extensive Games. International Journal of Game Theory 4(1), 25–55 (1975)
49. Simon, H.A.: Administrative Behavior: A Study of Decision-Making Processes in Administrative Organizations. Free Press, New york (1947)
50. Smith, R.: The Contract Net Protocol: High Level Communication and Control in a Distributed Problem Solver. IEEE Trans. on Comp. C-29(12), 1104–1113 (1980)
51. Stuhlmacher, A.F., Champagne, M.V.: The Impact of Time Pressure and Information on Negotiation Process and Decisions. Group Decision and Negotiation 9(6), 471–491 (2000)
52. Sycara, K.P.: Resolving Adversarial Conflicts: An Approach Integrating Case-Based and Analytic Methods. PhD thesis, Georgia Inst. of Techn., Atlanta, GA, USA (1987)
53. Tversky, A., Kahneman, D.: The Framing of Decisions and the Psychology of Choice. Science 211(4481), 453–458 (1981)
54. Tversky, A., Kahneman, D.: Rational Choice and the Framing of Decisions. Journal of Business 59(4), 251–278 (1986)
55. von Neumann, J.L., Morgenstern, O.: Theory of Games and Economic Behavior. Princeton University Press, Princeton (1944)
56. Weber, M.: Wirtschaft und Gesellschaft: Grundriss der verstehenden Soziologie, 1st edn. J. C. B. Mohr (Paul Siebeck), Tübingen (1922)
57. Woodward, J.: Management and Technology. Her Majesty's Stationery Office, London (1958)
58. Woodward, J.: Industrial Organisation: Theory and Practice. Oxford University Press, London (1965)
59. Wooldridge, M., Bussmann, S., Klosterberg, M.: Production sequencing as negotiation. In: Proc. of the 1st Int. Conf. on the Practical Application of Intelligent Agents and Multi-Agent Technology, London, UK, pp. 709–726 (1996)
60. Zlotkin, G., Rosenschein, J.S.: Negotiation and Task Sharing among Autonomous Agents in Cooperative Domains. In: Proc. of the 11th Int. Joint Conf. on Artificial Intelligence, San Mateo, CA, pp. 912–917. Morgan Kaufmann, San Francisco (1989)

Solution Architecture for Visitor Segmentation and Recommendation Generation in Real Time

Philip Limbeck[1] and Josef Schiefer[2]

[1] Senactive IT Dienstleistungs Gmbh
philip.limbeck@senactive.com
[2] Institute for Software Technology and Interactive Systems
Vienna University of Technology
js@ifs.tuwien.ac.at

Abstract. With increasing product portfolios of eCommerce companies it is getting harder for their customers to find the products they like best. A solution to this problem is to analyze the customer's behavior, and recommend products based on ratings. By considering click-stream data, the customer is unburdened with explicitly rating his favored. In this paper, we introduce a system for segmenting visitors and recommending adequate items in real time by using an event-based system called Sense and Respond Infrastructure (SARI) for processing click-stream data. We present the architecture and components for a real-time click-stream analysis which can be easily customized to business needs of domain experts and business users. SARI provides functionality to monitor visitor and customer behavior, respond accordingly and at the same time optimize and adapt customer processes in real time. To illustrate this approach, we introduce a reference implementation, its underlying infrastructure and business scenarios.

1 Introduction

Ever since, commerce was built highly upon personalized services, from the small grocery store to shoemakers; it has always been commonplace to be served and treated as an individual. But in the last century, industrialization and mass production started forcing people silently and slowly to give up this privilege and purchase goods without any personal appeal driven by salespeople. While business-to-business platforms already picked up the topic of individual treatment and online segmentation, many small traders and every-day online customer shops available today missed the opportunity to recognize individual analysis of their customers because of complexity and effort becoming high very fast. On the contrary, most web visitors are appealed by interspersing personalized content. By narrowing the mostly huge portfolio a web shop offers, the customers' time searching for specific items of interest is reduced drastically. There are basically two dimensions for measuring the goodness of recommendations: (1) quality – which is usually expressed as the average deviation between the recommendation and what the customer really looked for and (2) performance - the amount of visitors who can be served with recommendations at a given amount of time. Additionally, there are other things to consider, for instance the *cold-start problem*. A visitor impression about a given product can either be expressed explicitly by using questionnaires or ratings, or

G. Psaila and R. Wagner (Eds.): EC-Web 2008, LNCS 5183, pp. 102–113, 2008.

implicitly through mining data directly from visitor's behavior when searching and purchasing the product – including click-stream data from a website. There are at least three levels of data available with a different lifespan and expressiveness. It is necessary to model these levels differently to allow an efficient click-stream analysis. At the basic level, there is click information, which has no historic content whatsoever, but contains URL and navigation information. Above that is a layer containing all clicks made in one session and lying on top of that is a history layer which contains all available sessions for a specific user. To find out which sessions belong to the same user history is very challenging since client-sided cookie information can be removed very easily. To solve that problem, analysis is hooked upon the session, and subsequent sessions by the same user are evaluated separately.

Because it is necessary to produce recommendation data in real time, approaches like CEP (Complex Event Processing) introduced by Luckham [5] can be applied to monitor user behavior and react on sudden, unexpected changes as fast as possible. By modeling the user interactions with the front-end eCommerce platform as events, systems based on CEP can be used to mine important information about the visitor's behavior. Such event-based data, in combination with an historic backlog, can then be used to generate recommendations and other marketing and sales related tools in real time. By bridging available business processes with business intelligence using event streams, business knowledge can be generated from the stream data. This approach differs greatly from traditional systems which require data to be loaded and integrated in batch from different sources.

In this paper, we introduce a recommender solution for eCommerce platforms, based on the event-based system SARI [12], which is an event-based system capable of processing a large volume of events. The system detects business situations or exceptions and analyzes them automatically. Proactively or reactively, responses to situations like early warnings, preventing damage, loss or excessive cost, exploiting time-critical opportunities, or adapting business systems can be created.

In SARI, business processes and their exceptions are modeled using sense and re-spond decision trees and event-based scoring, which are both designed to be defined by business users. For rules triggered by events, SARI offers a user-friendly graphical interface which allows to model rules by combining conditions and their responses into an easy understandable tree structure [11].

2 Related Work

In the field of recommender systems, the major effort goes into optimizing Collabora-tive Filtering (CF) by combining it with aspects of knowledge-based systems, content-based filtering and statistical techniques to provide recommendation despite lacking data about a user. A lot of effort goes into dealing with the problems of spar-sity and making recommendation generation and their predecessor algorithms scalable enough to be deployed in large-scale eCommerce sites. The major problem with scal-ability here is updating the underlying model fast enough to provide accurate and real-time recommendations. Approaches range from algorithms for clustering the user or the item to reducing the dimension space using Latent Semantic Analysis (LSA), Principal Component Analysis (PCA) or Eigenvectors [3]. Hybrid variants use a weighted user-based and item-based generation of a neighborhood model. We overcome this obstacle by using an event-based approach which combines model and recommendation generation into one single step. A rule-based approach is presented in [1], which divides the analyzed meta-data of items into objective and subjective types, where objective types are described as direct properties of items, e.g. its author or album and subjective meta-data the user-specific properties rated individually, such as expressiveness and originality. Since those two types are treated using a combined approach of classical CF algorithms and rule-based evaluation of content-filtering using meta-data, the sparsity and the cold-start problem is subverted elegantly. However, neither site duration nor other forms of implicit data is analyzed and the model relies heavily on a necessary pre-rating step of training items.

Approaches in Web Usage Mining and user classification avoid asking for explicit ratings, but instead mine existing web-logs and navigational click-stream data. Approaches here include a probabilistic LSA model [4] and clustering as used by SUGGEST [2], which generates a co-occurrence matrix on-line. Association rules have been used by Mobasher et al. [6], producing frequent item sets which are later on fed into an on-line recommendation engine producing a set of recommended items. A taxonomy of the different approach types has been provided by Srivastava et al. [13], containing detailed information about the steps in web usage mining.

The underlying infrastructure of this solution architecture has been presented by Schiefer et al. [12]. They proposed SARI for event-driven applications on which the implementation of our visitor segmentation and recommendation generation system is based upon. Although SARI has not been applied to eCommerce applications yet, other applications exist, for instance in the field of fraud detection, where Rozsnyai et al. [9] presented a solution for detection of and reaction to fraud in the context of betting brokers.

3 System Architecture for Click-Stream Monitoring

In this paper, we introduce a solution for creating visitor segmentation and item-based recommendation creation using CF in real time. In this section, we give an overview of the system architecture. Furthermore, we discuss its underlying requirements to create recommendations and visitor segmentation for an eCommerce platform acting as front-end. Click-stream data are captured during the interactions of a user with an eCommerce platform. SARI is used to compute response events based on the user's behavioral history and his current interest. Both depend on a model using event-based scoring and decision-tree evaluation which will be presented in subsequent chapters.

Fig. 1 shows the communication process necessary to exploit SARI using calls to a web service, which is in charge of sending the generated results back to the eCommerce platform. As it can be seen, message queuing middleware and synchronization mechanisms are used to capture the resulting events from the SARI engine and to immediately transmit the result data back to the eCommerce platform. Because interaction with the web shop is based upon clicks, they have to be modeled as events in SARI. Fig. 1 describes how such a click is transmitted to the SARI system and

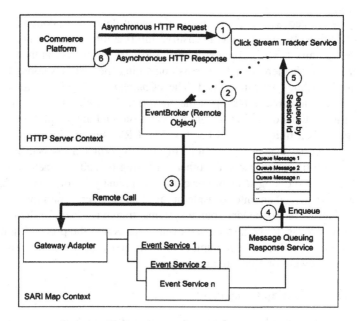

Fig. 1. Communication Process

how the result of that processing step is sent back to the front-end of the eCommerce platform. Through asynchronous HTTP channels, the data identifying the clicks is sent to the *Click Stream Tracker Service* (1) which then calls the *Event Broker* (2) to route the click data to a registered SARI map context with an adapter (3). The adapter processes this data and transforms it into a ClickEvent. After all operations inside the map context have finished, they sent a cumulative result to a special *Response Service* which enqueues the result event to a Message Queuing subsystem (4). Until now, the *Click Stream Tracker Service* has been waiting for the resulting information, and dequeues the data by the exploiting the unique user session identifier (5). Finally, the data is returned to the calling eCommerce platform (6), where the recommendations and other piggybacked data presented to the user.

4 Visitor Segmentation and Recommendation Generation with Event Processing Maps

4.1 Propagating Processing Maps with Events

Events are propagated through *event adapters (1)(2)* in Figure 2 from various source systems into the event processing maps. Propagated events that are processed by adapters are transformed into *event objects* that match corresponding *event type* definitions [11].

In our use case, there are two event adapters, where (1) is attached to the eCommerce web shop and (2) is attached to the customer relationship system of the eCommerce application. Since every event originating from the eCommerce platform must be uniquely associated with a given visitor, the event data has to contain its session identifier. Additionally, other context-sensitive information such as time and cookie data must be contained. The *eCommerce Web Shop Adapter*'s task is then to analyze those events and emit more specific events containing the visitor's context and type information, describing what happened. The *eCommerce Web Shop Adapter* (1) is attached to the eCommerce web shop of the platform. The web shop is used to implicitly track all the actions of the visitor on the web front end and display resulting recommendations, his affinity and CRM status. The CRM system, integrated using the *CRM Adapter* (2), is responsible for tracking the important user-specific measures concerning the profit, return quote and other marketing-related parameters. Since they can influence the processing steps of our click-stream analysis solution, CRM-related events must also be taken into consideration. The adapters continuously receive events from the two recommender source systems, transform them to their according event types and propagate the instantiated event objects into the processing map. The source systems can produce a large number of different event types, but only a part of these types are relevant for click-stream analysis.

4.2   Interconnecting Map Components

Every event service must have one or more input and output *ports* for receiving and emitting events. Each port must correspond to an event type.

In this use case, there are dozens of events emitted by the event adapters from the source systems, but only several are required for click-stream analysis using the EPM. The events ClickEvent and CRMInfo are routed to *event services* for processing.

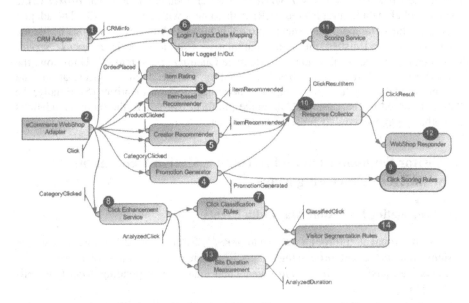

Fig. 2. Event Processing Map for Visitor Segmentation and Recommendation Generation

The rounded rectangles in Figure 2 represent *event services (3)(4)(5)(6)(8)(10)(12)(13)*, which are custom components for processing events. The first three events services are used to create individual recommendations and promotions based upon an affinity model of users and items. The event services (8), (10) and (12) are infrastructural services, responsible for data staging and resynchronization and (13) is in charge of measuring times between clicks.

4.3 Event Services

The event services' *Item Rating* and *Item-based Recommender* (3) job is to process incoming orders and products to create a model for CF. The *Item Rating* event service has to rate a set of items contained in an OrderPlaced event using different nearest-neighbourhood (NN) functions[10] and transform the result to a discrete value which is then stored to model the affinity between all the items in the OrderPlaced event and all other items in the customer's order history. The *Item Rating* event service accomplishes this task by sending scoring events to the nearby *Scoring Service* (11). Based on that information, the event service's *Item-based Recommender* task is to create recommendations based on the order history of the known customer. It creates a set of top N matching products related to the product ID of the ProductClicked event. It accesses the data which has been generated previously by the *ItemRating* event service. The resulting recommendations are published as ItemsRecommended event.

The *Promotion Generator* (4) is responsible for creating real-time promotions based on the current classification of the user. There are multiple types of classifications which correspond to different category levels on the front-end side. Promotions consist of categories corresponding to the cross-product of all available classifications and their restrictions opposed to the items and categories in the front-end. Furthermore, the CustomerId of the CategoryEvent is considered to treat customers differently according to CRM attributes such as return value or profit, which results in different heights of benefit or other parameters strong related to promotions.

The response service *Web Shop Responder* (12) has the special task of enqueuing the resulted events in the message queuing system which acts as middleware between the Web Shop and the SARI system. In general, response services are considered to communicate generated results towards the outside world.

Since there are multiple results to consider from different branches of the SARI map, for instance the branch of the *Promotion Generator* (4) may send a ClickResultItem as well as the *Item-based Recommender* (3); those events must be rejoined and sent back to the requesting eCommerce front-end at the same moment. To overcome this problem the *Response Collector* (10) service uses a correlation set which correlates over all ClickResultItem events that contain the same unique PublishingId. If all events arrive, a bulk ClickResultEvent, containing all ClickResultItem events, will be emitted.

The *Click Enhancement Service* (8) communicates with the external database where product, category and classification mapping information are stored. After the event data is enhanced, the incoming CategoryEvent is disaggregated according to its Categories attribute into AnalyzedClick events. The Categories attribute contains a path in a hierarchy from the top-most category down to a single product, uniquely

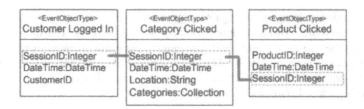

Fig. 3. Correlation Set

identifying the current location inside the eCommerce front-end. The CategoryEvent also contains additional meta-information such as the session identifier and – if CRM data is available about the user – a CustomerId.

The event service *Site Duration Measurement* (13) processes AnalyzedClick events. Whenever such an event is received, a timer is started for the level of the category path the AnalyzedClick event belongs to. If in the meantime, a second AnalyzedClick of the same user and the same category path level arrives, an AnalyzedDuration event is published, which contains the time span between the two AnalyzedClick events. If the started timer runs out, a stop event will be published, having the maximum time span of the specific level.

The *Creator Recommender*, which is represented by (5) in Figure 2, also has the ability to create recommendations. But instead of an analysis using CF techniques, cross-category browsing is analyzed, to find preferences concerning a specified creator – which can be an author or artist of a product item. Every time a ProductEvent comes in, the creator of that product is stored for later use and kept until a CategoryEvent appears. For all products, the most occurring creator is ranked due to weighted relative occurrence. For the top N products found, ItemRecommended events are generated containing products with the most-preferred creators and the current category the user is visiting at the moment.

Data Mapping Login/Logout (6) is a special event service which is responsible for mapping incoming events of type UserLoginLogout to a database, which stores the login history of all users. Additionally, all the resulting affinity groups a user belongs to during a session in front of the eCommerce platform are also stored.

4.4 Correlating Events

An important concept that is used in our solution is event correlation. For purposes of maintaining information about business activities, events capture attributes about the context, when the event occurred. Event attributes are items such as the agents, resources, and data associated with an event, the tangible result of an action (e.g., the result of a transport decision), or any other information that gives character to the specific occurrence of that type of event.

Elements of an event context can be used to define a relationship with other events. Components in the event processing map use correlation sets for defining relationships between events of business activities. Sense and respond rules use correlation sets to collect attribute values of correlated events. The gathered event data is held in a data container which we define as a correlation session. For a detailed discussion on the management of correlation sessions, please refer to [7].

The event service *Creator Recommender* makes use of event correlations (Fig. 3) to collect clicks from the same user identified by his session identifier. In other words, there is a correlation that brings together all events that have the same SessionID. That allows services to track clicks coming from the same visitor and correlating the SessionID with an already existing CustomerID which identifies a customer in the applications CRM system.

4.5 Scoring Service

The *Scoring Service* (11) in Figure 2 is a service that receives so-called *scoring events*. Basically, there are two main event types that are consumed by the Scoring Service. An *IncrementScore* lets the service increment the score value and a *DecrementScore* that decrements the specified score. Additionally *ResetScore* lets the service reset the score value to its default value determined by its *ScoreType*. A scoring event contains following attributes:

1. A *ScoreType* that is an identifier for the score.
2. A *Classifier* that can be used to relate a score to a user by its Session ID
3. A *Sub Classifier* that can be used to create sub category for a score.
4. A *Value* that increments or decrements the specified score corresponding to the classifier and sub classifier.

Scoring can be used to model a certain state for the Event Processing Map as this would be unachievable otherwise without using the previously introduced concept of event correlation.

The affinity among items which is the result of Item-based CF can be modelled using scoring. The result of a CF algorithm for instance Pearson Nearest Neighbourhood can be mapped to a discrete value which is then represented by the value of the given score. Depending on this ItemSimilarity score, recommendations can be generated if their similarity reaches beyond a preset threshold value and/or is one of the top N available scores. Another usage is to assign the visitor to a preset segment which allows grouping of visitors into similar affinity classes. For instance, visitors with a high duration on books related to cooking in combination with a high click count of book related topics will be assigned to a segment which describes his affinity to books and cooking at best. Since all that data is modelled using score values, there is a global state available for each user which can be accessed by each map component easily.

4.6 Rule Service

The rectangles (7)(9)(14) in Figure 2 are *rule services* containing business rules to steer the process flow. The rules are comparable to decision trees to a limited extend and can be customized by the user through a graphical user interface. Like hubs, event services and rule services have ports that can be restricted to event types. Behind the rule services in event processing maps, there is a sophisticated rule management system [11]. SARI allows it's users to model business situations and exceptions with *sense and respond rules* that have been designed to be maintained and created by business users and domain experts in first place. Sense and respond rules allow to

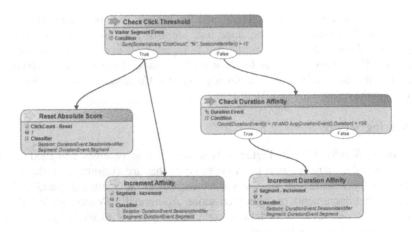

Fig. 4. Rule Example

describe and discover business situations and can automatically trigger responses such as generating early warnings, preventing damage, loss or excessive cost, exploiting time-critical business opportunities, or adapting business systems with minimal latency.

Sense and respond rules are organized in rule sets and allow constructing decision scenarios, which use event conditions and event patterns for triggering response events. Fig. 4 shows a rule taken from the *Visitor Segmentation Rules* (14) service in Figure 2 for explanatory purposes.

Event conditions and event patterns can be arbitrarily combined with logical operators in order to model complex situations. Response events are generated when event conditions or event patterns evaluate to true. EPMs allow to model which events should be processed by the rule service and how the response events should be forwarded to other event services. *Event conditions* are the rectangles with 'True' and 'False' ports in Fig. 4. Event conditions have a triggering event and a list of Boolean expressions that are evaluated when the triggering event is received. In this case, the condition *Check Click Threshold* is triggered when the event VisitorSegmented is received by the rule service. In such a case, the service checks if the score Click-Count, with the visitor's given SessionID, is greater than 10. If the condition evaluates to true or false the corresponding branches will be evaluated further until a response event is reached.

Scoring Response events, the remaining rectangles without any ports in Fig. 4, are one type of actions that is initiated when event conditions evaluate to true and/or event patterns match. Response events can be arbitrary events with attributes that can be bound to attributes of triggering events. In the example, in Fig. 4 the first condition *Check Click Threshold* evaluates to true and the special scoring response events *Increment Affinity* and *Reset Absolute Count* are triggered. Those scoring response events differ from usual response events in a way that they enable direct modification of the scoring model presented in section 4.5. Both response events are Scoring event types with the attributes set in the attribute binding fields.

The event processing map in Figure 2 contains three rule services: The task of the Click Scoring service (9) is to determine the type of the incoming click and react by changing the visitors short-term affinity towards a given customer segment. In other words a weighted model is created which can be used by other event services to infer promotions for the currently most affine profile. Overriding rules for example to treat the first contact with a specific class are also implemented here. The role of the Click Classification Rules (7) is to watch the pre-built short-term model for a specified count of clicks to be reached. When this condition is triggered for a visitor, its short-term classification is reset and his long-term affinity classes incremented according to the relative occurrence of the collected classes.

5 Implementation

To demonstrate the presented approach, we integrated Senactive InTime™, which is an implementation of SARI, with OsCommerce, a free eCommerce platform providing simple web shop functionality. Figure 5 shows a scene from the live demo available online at http://www.intelligentoffer.net. The left screenshot shows the categories matching the current visitor segment, while the right one shows three products which have been generated by the item-based recommender engine. Visitor segments are per default modeled according to the hierarchy of the categories, but user-defined segments containing other categories and items are possible as well. We implemented the Item-based Recommender service (3) from Figure 2 using a rather simple NN algorithm based on the presented Scoring model, but the flexibility of the presented solution allows more complex algorithms to be integrated easily. Since OsCommerce is based on PHP and MySQL, the appropriate infrastructure is needed.

We used the Microsoft Internet Information Services (IIS) with a corresponding PHP plug-in for running OsCommerce which in turn uses MySQL to store product

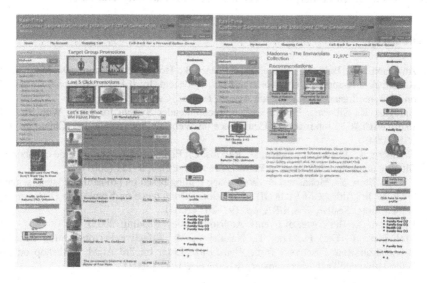

Fig. 5. OsCommerce Webshop using the presented solution

data. To infer visitor segmentation and generate recommendations, essential subjective meta-data data is separately stored on a Microsoft™ SQL Server 2005 database needed by Senactive InTime™. An ASP.NET web service is running inside IIS to tap the click information generated by the visitor and send it to the incoming adapters of Senactive InTime™ using Microsoft™ Message Queuing. Providing just an API for sending click-stream data and receiving a processed result has the advantage of not having to modify big parts of the platform as this may result in cumbersome changes to be made when the eCommerce platform is difficult to adapt.

6 Conclusion and Future Work

In the future, recommender systems will gain even more relevance than they have today. Also personalized services focus an ever increasing attention by eCommerce vendors. There are still problems to be tackled especially concerning scalability for large-scale eCommerce websites and quality of recommendations. The proposed solution is able to solve this problem and manages to do the decisive step which assists the Internet customer of today with finding the products he likes best. Among these solutions, this paper contributes by presenting the architecture and necessary infrastructure for click-stream analysis using an event-based approach. Our major advantage compared to other solutions is the seamless integration of any available click-stream data implicitly, allowing us to merge the steps of creating a model for item and user correlation, and generating recommendations out of the resulting data. Additionally site duration is also considered as it is an important indicator for interest besides user clicks. In the future we plan to tackle basically two aspects: (1) improvement of performance and scalability to increase relevancy for eCommerce operators with a larger customer and product set using clustered item-item CF, and (2) experimentation with other, more recent algorithms currently available.

References

1. Anderson, M., Ball, M., Boley, H., Greene, S., Howse, N., Lemire, D., Mcgrath, S.: Racofi: A rule-applying collaborative filtering system. In: IEEE International Conference on Web Intelligence/Collaboration Agents, Halifax, Canada (2003)
2. Baraglia, R., Silvestri, F.: An Online Recommender System for Large Web Sites. In: IEEE/WIC/ACM International Conference on Web Intelligence, Beijing, China, pp. 199–205 (2004)
3. Goldberg, K., Roeder, T., Gupta, D., Perkins, C.: Eigentaste: A constant time collaborative filtering algorithm. Information Retrieval Journal 4(2), 133–151 (2001)
4. Jin, X., Zhou, Y., Mobasher, B.: Web usage mining based on probabilistic latent semantic analysis. In: 10th ACM SIGKDD international Conference on Knowledge Discovery and Data Mining, Seattle, WA, USA, pp. 197–205 (2004)
5. Luckham, D.: The Power of Events. Addison Wesley, Reading (2005)
6. Mobasher, B., Dai, H., Luo, T., Nakagawa, M.: Effective Personalization Based on Association Rule Discovery from Web Usage Data. In: 3rd ACM Workshop on Web Information and Data Management, Atlanta, GA, USA (2001)

7. Rozsnyai, S., Schiefer, J., Schatten, A.: Event Cloud - Searching for Correlated Business Events. In: 9th IEEE Conference on E-Commerce Technology and the 4th IEEE Conference on Enterprise Computing, E-Commerce and E-Services, Tokyo, Japan (2007)

8. Rozsnyai, S., Schiefer, J., Schatten, A.: Concepts and Models for Typing Events for Event-Based Systems. In: Inaugural International Conference on Distributed event-based systems, Toronto, Canada (2007)

9. Rozsnyai, S., Schiefer, J., Schatten, A.: Solution architecture for detecting and preventing fraud in real time. In: 2nd International Conference on Digital Information Management, Lyon, France, vol. 1, pp. 152–158 (2007)

10. Sarwar, B.M., Karypis, G., Konstan, J.A., Reidl, J.: Item-based collaborative filtering recommendation algorithms. In: 10th International World Wide Web Conference, Hongkong, China, pp. 285–295 (2001)

11. Schiefer, J., Rozsnyai, S., Schatten, A.: Event-Driven Rules for Sensing and Responding to Business Situations. In: Inaugural International Conference on Distributed event-based systems, Toronto, Canada (2007)

12. Schiefer, J., Seufert, A.: Management and Controlling of Time-Sensitive Business Processes with Sense & Respond. In: International Conference on Computational Intelligence for Modelling Control and Automation, Vienna (2005)

13. Srivastava, J., Cooley, R., Deshpande, M., Tan, P.-T.: Web Usage Mining: Discovery and Applications of Usage Patterns from Web Data. ACM Special Interest Group on Knowledge Discovery and Data Mining Explorations 2(1) (2000)

Effects of Cultural Background on Internet Buying Behaviour: Towards a Virtual Global Village?[*]

Efthymios Constantinides[1], Carlota Lorenzo[2], Miguel Ángel Gómez-Borja[2], and Peter Geurts[3]

[1] University of Twente. School of Management and Governance, Department NIKOS
P.O. Box 217. 7500 AE Enschede, The Netherlands
E.Constantinides@utwente.nl
[2] University of Castilla-La Mancha. Faculty of Business. Plaza de la Universidad, 1. 02071, Albacete, Spain
Carlota.Lorenzo@uclm.es, MiguelAngel.GBorja@uclm.es
[3] University of Twente. School of Management and Governance, Departement of Research Methods and Statistics. P.O. Box 217. 7500 AE Enschede, The Netherlands
P.A.T.M.Geurts@utwente.nl

Abstract. The article identifies the combined influence of the Web Experience (or online experience) components on the online consumer's behaviour from three different angles: (1) The relative importance of the online experience factors in choosing an online vendor, (2) the actual effect of these factors on the vendor choice and (3) the influence of personal and behavioural characteristics on the virtual shopping behaviour. The results identify early symptoms of an emerging behavioural convergence among Internet users of different cultures and nationalities, suggesting that cultural and behavioural differences in the physical world could have limited influence on people's behaviour in the virtual marketplace. This outcome could suggest the emergence of a global virtual village, an issue worth of further scholastic research but also an issue of particular importance for global Web vendors and web site designers.

Keywords: Web experience, Internet Marketing, cultural differences, buying behaviour, Web Marketing, marketing strategy.

1 Introduction

The Web Experience has been defined as a four stage process describing the successive steps of an online transaction [1]: Encountering the online retailer's home page, selecting a product from the online catalog, completing the order form and accessing customer service and support. The Web Experience has been also described as "the consumer's whole perception about the online company resulting from his/her exposure to a combination of virtual marketing tools" [2]. This perception is likely to have a substantial and direct effect on the buying behaviour of the online consumer. It embraces several elements, under the control of the online marketer, affecting the process of searching, browsing, finding, selecting, comparing and evaluating information as well

---

[*] This study was developed within the Research Projects whose references are PCI08-0004 (JCCM, 2008-2010) and TC20070056 (UCLM, 2008).

G. Psaila and R. Wagner (Eds.): EC-Web 2008, LNCS 5183, pp. 114–123, 2008.

as interacting and transacting with the online firm [3]. The online customer's perceptions and actions are influenced by design, events, emotions, atmosphere and other visual or psychological cues but also by the web site's usability and the information becoming available during interaction. There is plenty of recent research about such - controllable by the online marketer - factors influencing the online customer's behaviour and the way these factors contribute to the online experience. [4] proposed a taxonomy summarizing the findings of studies published in 48 research papers between 1997 and 2003. This taxonomy identifies five groups of parameters (Usability, Interactivity, Trust, Aesthetics and Marketing Mix) as dimensions of the controllable Web Experience. There is plenty of evidence that the large majority of wired consumers consider by now the web as their primary source of information when searching for products, services, news, weather, travel directions or entertainment: according to a recent BurstMedia survey 57.1% of all US web users over 18 years old use the Internet as their primary source of information about products and services they intend to buy[1]. This percentage is even higher (69.2%) among the most affluent consumers, with income of $75.000 or more. In the background of these developments it is not surprising that marketers increase their efforts to attract audiences to their web sites, something evident in the substantial increase of online marketing budgets.

The main objective of this paper is to analyze the impact of web experience on virtual buying behaviour. Based on the background literature as a starting point, we propose a model to study the significance of the constructs analyzed. Moreover, a number of personal variables have been included in the model in order to observe their influence on Web Experience.

2 Research and Hypotheses

Research on the impact of WE-related factors on the online consumer's behavior far has been conducted mainly within culturally homogenous markets; cross-cultural comparison of the actual online behavior is an issue that has so far received limited attention. Considering however the global reach and effects of the Internet it is interesting to examine whether the cultural background of online consumers has any influence on the way the Web Experience factors affect online buying behavior. A rule of thumb followed by most global corporations operating web sites in several countries or geographical areas is that full customization of national web sites is the proper policy to follow. Yet there is no concrete evidence so far that this is the right strategy. International Marketing text books, based on research findings from comparison of consumer behavior in different cultures [5] place special attention to inter-cultural differences and their impact on international marketing strategies [6; 7; 8; 9]. While markets become global and more transparent, a variety of new Information and Communication Technologies (ICT) are getting increasingly accessible to consumers and producers all over the world. The issue of cultural diversity has entered a new round of academic debate focused on the effects of widely available ICT on the global consumer. This is by no means a new issue; Marshall McLuhan looking to the influence and spreading of the mass electronic media worldwide proposed the concept of Global Village back in 1964 [10]. Several academics and practitioners have embraced

---

[1] MarketingVox, April 20, 2006.

the global village notion arguing that globalization and new technologies has lead to cultural convergence and more homogenous consumer behavior patterns across countries and cultures. [11] and [12] reject the notion of cultural convergence arguing that disparities in consumption patterns in different countries have not been reduced as a result of new technologies and therefore cultural differences should be given special attention in designing global market entry strategies. Most attention in this dispute is focused on convergence at consumption patterns on macro or micro level but limited attention has been paid to the question whether the global adoption of new ICTs and most notably the adoption of the Internet, have lead to convergence or divergence of the actual consumer buying behavior patterns at a global scale. The objective of this study is to address this issue by studying the online consumer behavior in two European countries with different and distinct cultural background, namely Spain and Netherlands. In that respect the following hypotheses are formulated:

H1: The influence of the Web Experience elements (a: Usability, b: Interactivity, c: Trust, d: Aesthetics, e: Marketing Mix) on online shopping behavior of Spanish and Dutch consumer is similar.

Personal consumer attributes (involvement, motivation, experience, ability to Internet adaptation, and so on) affect the purchase process and final decision [13; 14]. [15] included three personal variables (i.e. the motives to shop online, the level of experience in Internet use and online shopping) in their online behavior model. Regarding the motivation variable, they observed neither influence of shopping motives on the way people experience the different web factors nor the choice of a virtual vendor. However, they observed a positive influence of the experience as personal variable. Based on that work and the literature supporting the argument of cultural convergence and homogenous consumer behavior patterns across countries and in order to analyze the influence of these two specific internal variables (i.e. motivation and experience –years of Internet experience and familiarity with online shopping, according to [15]–) on users' preferences, the following hypotheses are proposed:

H2: The online buying behavior of Spanish and Dutch consumers is not influenced by online shopping motivations (a: Motives towards online shopping) and likewise nor influenced by personal characteristics (b: Years of Internet experience; c: Familiarity with online shopping).

3 Method

The study was conducted by means of an identical survey among undergraduate students at two Universities in Spain (204 students) and The Netherlands (85 students) based on simulating a virtual shopping activity in the computer lab. Subjects were instructed to carry out a specific online shopping task: searching for information, finding and "buying" a digital camera meeting a number of technical and economic specifications (e.g. up to a fictitious budget of 300 Euros including price and postal costs). After completing the task participants were asked to fill in an online questionnaire. The instructions were explained at the beginning of the experiment by supervisors. The conditions regarding searching, choosing the vendor and the camera brand and type were as realistic as possible; the participants were free to buy the product in any online shop anywhere in the world provided that this was able to be delivered in their country. There were no limitations on the number of web sites visits or search methods used. The camera had to be new and the total time available was 90 minutes; this time was sufficient as pre-testing indicated. During the shopping, subjects had to create two lists of vendors: "Favorites" and "No favorites" according to their experience during the interaction with the online shops and were asked to complete the online ordering procedure but stop before confirming the purchase. The survey was divided in two parts. The first part (Questionnaire I) was meant to identify the participants' demographics and attitudes towards online shopping, their previous experience with the Internet (buyers/no buyers) and their main motives for shopping or not shopping online. Once this step was completed, each participant had to proceed with the online shopping activity and fill in the rest of the questionnaire (i.e. Questionnaires A, B, and C). The form A was about the Web site they "bought" the camera), form B was about the web site from the top of the Favorites list. List C was about the web site where participants would never buy the camera from, for whatever reason. The lists A, B and C were identical. Participants had to indicate whether they agree or disagree with statements related to their experience from interacting with each of these three websites on 25 individual features relevant to the five Web Experience factors identified earlier (e.g. "It is convenient to buy products in this online shop", "the shop offers excellent customer service", "the site offers adequate guarantees for the safety of online transactions", "the site's design is superb", "the site offers a wide deep product assortment"...). The answers were given in a five-point Likert scale with values ranging from 1 (fully disagree) to 5 (fully agree). The surveys and the whole process were tested with preliminary surveys, meant to identify potential problems in the scenario and the questions. The dependent variable related to online shopping behavior was measured as dummy variable (i.e. buy or not buying) which indicates the consumer' shopping preferences towards virtual vendor.

4   Descriptive and Statistical Results

A comparative overview for the main descriptive characteristics for the samples in both countries is presented in table 1.

Table 1. Demographic and Personal characteristics of survey participants

	Spain	**The Netherlands**
Gender	Female: 63% Male: 37%	Female: 29% Male: 71%
Age	18-22	20-25
Experience with the Internet (more than four years)	34%	84%
Purchasing using the Internet	31%	77%
Amount spent yearly (between 50-100€)	15%	21%

Table 2. Consumer's purchase preference from web experience factors

Dependent variable (buy/not buying)		Consumer Preference	
Hypotheses		H1 (a, b, c, d, e)	
Countries comparison		SPAIN	NETHERLANDS
Nagelkerke		.34	.31
Hoshmer Lemeshow		14.99 (8)	48.89 (8)
WE Factors – H1 (Independent variables)	a: Usability	1.29 (.15)*	.74 (.22)*
	b: Interactivity	.19 (.13)	-.27 (.20)
	c: Trust	.55 (.11)*	.24 (.18)
	d: Aesthetics	.47 (.12)*	.11 (.20)
	e: Mk Mix	.54 (.13)*	.55 (.22)*

Legend:
· Proxies of standardized regression parameters are presented in the cells.
· Asterisk (*) indicates statistical significance on 5% level (i.e. confidence level: 95%).
· Between brackets are the standard errors.

Table 3. Influence of personal characteristics on consumer's purchase preference

Dependent variable (buy/not buying)		Consumer Preference			
Hypotheses		H2a		H2b,c	
Countries comparison		SP	NL	SP	NL
Nagelkerke		.46	.33	.35	.48
Hoshmer Lemeshow		7.95 (8)	34.57 (8)	9.80 (8)	11.06 (8)
WE Factors (Independent variables)	Usability	1.82 (.37)*	.74 (.22)*	1.34 (.15)*	1.07 (.26)*
	Interactivity	-.56 (.28)	-.28 (.20)	.18 (.13)	-.36 (.21)
	Trust	.96 (.21)*	.23 (.18)	.58 (.12)*	.35 (.20)
	Aesthetics	.72 (.26)*	.12 (.20)	.47 (.12)*	.16 (.22)
	Mk Mix	.71 (.25)*	.56 (.21)*	.55 (.13)*	.72 (.25)*
Main Motive – H2 (Independent variable)	a: To find better prices **	-.15 (.40)	-.76 (.39)*		
Experience – H2 (Independent variable)	b: Years of Internet usage			-.11 (.07)	-.26 (.09)*
	c: Familiarity with online shopping			-.18 (.22)	.11 (.31)

Legend:
· Proxies of standardized regression parameters are presented in the cells.
· Asterisk (*) indicates statistical significance on 5% level (i.e. confidence level: 95%).
· Between brackets are the standard errors.
· Double asterisk (**) indicates that "to find better prices" is the motive to buy online with the highest impact. So, the rest of the identified motives are not reported in this table. In both countries that variable has been the most selected reason by participants.

The three websites chosen by users in the lists A, B and C were analyzed on the basis of the responses on 25 statements per web site describing the participant's perception on various WE characteristics. In order to test the hypotheses, a factor analysis was carried out to reduce the number of WE items. A binomial logistic regression was executed with the five WE factors per website as independent variables and the purchasing behavior as the dichotomy explained variable (i.e. buy/not buying). The users' buying preferences regarding the WE elements and the influence of their personal characteristics can be seen in tables 2 and 3.

Based on the results of the previous analysis the hypotheses proposed were evaluated. The arguments are showed in the following paragraphs.

H1: The influence of the Web Experience elements (a: Usability, b: Interactivity, c: Trust, d: Aesthetics, e: Marketing Mix) on online shopping behaviour of Spanish and Dutch consumer is similar

A comparative analysis between both countries shows differences and similarities on the importance of WE elements on consumer decisions in the choice of an e-vendor (table 2). The **Usability** and **Marketing Mix** elements have a positive and significant influence on consumer decisions in both countries. Therefore, H1a and H1e are not rejected. Regarding this point, it is important to notice the behavior of statistical coefficients –beta parameters– in both cases. Specifically, regarding the Marketing Mix factor, the behavior in both parameters is similar (Spanish .54 versus Dutch .55). The analysis of the Spanish data indicates that the Usability factor has a higher influence than in the case of the Dutch data (i.e. Spanish 1.29 versus Dutch .74). These results indicate that the usability elements are more relevant for the Spanish population, probably due to a lower experience with the Internet than in the case of Dutch users (table 2). This could suggest that Spanish people appreciate more the easy navigation through web sites. As to the **Trust** and **Aesthetics** factors, these are significant influencers in the case of the Spanish study but no significant in the case of Dutch consumers; H1c and H1d should be rejected because the effect of these factors on the two groups is quite dissimilar. According to descriptive data (table 2), the majority of Spanish participants have no much previous experience in the use of the Internet and online purchase whereas for Dutch participants the percentages are higher (i.e. 84 versus 34, and 77 versus 31, respectively). Consequently, we can conclude that Web Experience elements like Aesthetics and Trust are considered as less relevant by more experienced web shoppers (i.e. Dutch consumers). Subjects less skilled in the use and purchase on the Internet need more positive cues and information to be provided by online stores –such as high presentation quality, good atmosphere, safety in e-transactions, transparent guarantee policy, and so on– in order to choose an e-vendor. The more cautious behavior of the Spanish consumer is in line with findings about cultural behavior: According to Hofstede's cultural dimensions classification [16; 17; 18] the "uncertainty avoidance" variable is higher in Spain. This could explain the need of having more web stimulus and patterns for efficiently completing the purchase task in case of Spanish online users. Additionally, in both cases (Trust and Aesthetics factors), the regression parameters are higher in the Spanish study (Trust: .55 versus .24; Aesthetics: .47 versus .11). This fact additionally confirms the explanation about the higher impact of both of these Web Experience factors on the Spanish sample. Regarding the **Interactivity**, in both countries the effect of this factor on the choice of an online vendor is limited. Moreover there are no significant differences between the two populations; according to regression parameters, in both cases the Interactivity factor is not significant for the e-vendor choice. On the basis of this analysis the H1b is not rejected. In this case, according to statistical coefficients (table 2) it is interesting to point out the difference of the values of that factor between the two populations. In the Spanish case we can find a direct relationship between Interactivity elements and consumer's preferences while in the Dutch study the relationship is indirect (.19 versus -.27). These results indicate the low impact that web elements related to Interactivity factor (possibility to interact with personnel or other users buying in the online store, etc.) have on Spanish users. In the case of Dutch users, the effect of that factor is negative leading to the interesting conclusion that these shoppers are choosing web sites that are not interactive in the sense described. These effects are again in line with Hofstede studies [16; 18] indicating that the "individualism" cultural dimension in The Netherlands is higher than in Spain. Furthermore, unlike the rest of WE factors –specifically, Trust and Aesthetics–, more experienced Internet users seem to consider the inclusion of web elements allowing for interaction with the shop's staff as unnecessary and consequently they prefer that the stores put more attention to other web characteristics and tools directly related to Usability and Marketing Mix. According to the above results it can be argued that differences between distinctive cultural groups in the physical world are reflected in some aspects of the behavior of these groups as online consumers. Yet the differences are much smaller than one would expect on the basis of the Hofstede framework of cultural dimensions and probably have to some degree to do with the differences in experience of the Internet as buying channel. This could lead to the tentative conclusion (requiring further research) that the global virtual community tends to establish similar patterns of reaction to Web Experience elements and similar behavioral patterns in searching and purchasing through the Internet.

H2: The online buying behavior of Spanish and Dutch consumers is not influenced by motivation for engaging in online shopping and also nor influenced by personal characteristics (Years of Internet experience and Familiarity with online shopping).

H2a: Motives towards online shopping of Spanish and Dutch consumers will have the same effects on their online buying behavior.

As for as the "motives towards online shopping" (table 3), the variable related to "to find better prices" was chosen by the majority of participants in both studies as the main reason for shopping online. Yet the inclusion of this variable in both models, Spanish and Dutch had no substantial effect on way people make online consumer decisions. This fact indicates that actual behavior is not similar to the perceived one. Looking more closely to the actual results, we observe some important differences. In the case of Spanish study, the motivation factor is not significant within the model but improves the results if we compare the five factors before and after its inclusion in the model (see statistical coefficients of WE factors in tables 2 and 3). On the other hand, in case of Dutch study, although the motivation variable has a significant effect within the model, it does not produce any improvement to WE results. So, in both cases its effect is nil. As a result, the H2a is not rejected.

H2b: Years of Internet experience of Spanish and Dutch consumers will have the same effects on their online buying behavior.

H2c: Familiarity with online shopping of Spanish and Dutch consumers will have the same effects on their online buying behavior.

Regarding the independent variables "years of Internet usage" and "familiarity with online shopping" these show dissimilar results between both countries (table 3). In the Spanish case, both variables have no significant direct or indirect effect on the appraisal of the WE factors. Moreover, the differences between the beta coefficients before and after their inclusion in the model are not considerable (see comparison between statistical coefficients in tables 2 and 3). In the Dutch study, the effect of "familiarity with online shopping" has not significant effect on WE factors. The "years of Internet usage" as experiential variable is significant and has a higher indirect effect than in the Spanish study. In the Dutch study the experience with the Internet diminishes the role of web elements in online retailing. In fact, the inclusion of these internal variables in the model improves the results in all WE factors mainly in the Usability and Marketing Mix elements but in a negative direction (beta coefficient -0,26). According to the descriptive data (table 1) the experience with the Internet use by Spanish participants is lower than in the Dutch case (i.e. 34% versus 84% with more of four years of experience). In that context, the H2b and H2c could not be accepted. In fact, in the Dutch population we observe that the longer people have been using the Internet and the longer they shop in virtual stores, the more critical they become (i.e. they are more difficult to be satisfied) towards the Usability and Marketing Mix elements of web sites, while these personal variables do not show significant effect on the appreciation of the other three factors (i.e. Interactivity, Trust, and Aesthetics). In this sense, it is also relevant to mention that the "uncertainty avoidance" cultural dimension [17; 18] could exercise some influence in the shopper's behavior: Spanish people prefer to avoid the uncertainty of online shopping. Furthermore, according to the indirect effect shown in the data (table 3), the Spanish consumer prefers to be more intensively exposed to web experience elements in the online stores in order to carry out a purchase more efficiently. Finally, regarding the general variable called "experience" (i.e. years Internet usage and familiarity with online purchase) we can notice a different behavior in each study. In case of the Spanish study, the "familiarity with online purchase" has a more substantial impact on the consumer's preferences in choosing a virtual vendor than the "years of Internet usage", but the inclusion of this general variable does not improve the result model. In contrast, in Dutch study, the variable with more influence on e-vendor choice is the "years of Internet usage". In both studies, the general variable "experience" affects the consumers' preferences in a negative way: More experienced users with the Internet consider that more important web experience elements are the Usability and Marketing Mix and in the second place, Trust, Aesthetics, and finally, Interactivity. Based on this point, the H2b and H2c could be not rejected.

5   Conclusions and Issues for Further Research

This paper presents a comparative study on the influence of Web Experience factors on consumer's buying behavior in two countries with different cultural backgrounds:

Spain and The Netherlands. The results indicate that in both countries the factor Inter-activity does not play any substantial role on the choice of an e-vendor. The Spanish study shows that four of the five WE elements clearly influence the online shopper's preferences, in line with previous findings. The Dutch survey indicates that only two of the WE factors (Usability and Marketing Mix) have a substantial effect on online shoppers' preferences while the effect of Trust and Aesthetics is much more limited than the effects of these factors on the Spanish consumer. A possible explanation with regard to the different effect of the factors Trust and Aesthetics between for the two markets could be cultural and behavioral differences between consumers. Dutch con-sumers are more individualistic and characterized by less risk aversion in comparison to Spanish consumers [17; 18]. With regard to online shopping motivation, years of Internet usage and experience with online shopping, these personal factors do not seem to affect the online buying behavior of either the Dutch or the Spanish consum-ers. The study underlines a number of interesting issues as a basis for further research into inter-cultural differences in virtual buying behavior. Expanding the research to more European and International cultures is necessary considering the increasing significance of the Internet as a global marketing platform. Online vendors and web designers should be aware of the effects of cultural differences when designing global virtual stores for different cultural segments and different customers (e.g. different languages, different products, etc.). In conclusion, the study identifies some differ-ences but also a lot of similarities in the way online experience factors affect e-users from Spain and The Netherlands. One could expect that increasing globalization and adoption of the Internet worldwide will decrease rather than increase these differences in the future. Further research is necessary for defining the different effects of Web Experience on the behavior of online buyers of other types of products where image or design is very important (clothes, apparel, autos etc) customized products or intan-gibles on cross-cultural contexts.

References

1. Tamimi, N., Rajan, M., Sebastianelli, R.: The state of online retailing. Internet research, Applications and Policy 13(3), 146–155 (2003)
2. Watchfire Whitepaper Series: Bad things shouldn't happen to good web sites, best prac-tices for managing the web experience (2000),   ,
   http://www.watchfire.com/resources/search-and-ye-shall-find.pdf
3. Constantinides, E.: The 4s web-marketing mix model, e-commerce research and applica-tions, vol. 1(1), pp. 57–76. Elsevier Science, Amsterdam (2002)
4. Constantinides, E.: Influencing the online consumer's behavior: the web experience, jour-nal of internet research. Electronic Networking Applications and Policy 14(2), 111–126 (2004)
5. Douglas, S., Craig, C.S.: The changing dynamic of consumer behavior: implications for cross-cultural research. International Journal of Research in Marketing 14(4), 379–395 (1997)
6. Cateora, P., Hess, J.: International Marketing, 4th edn. Richard D Irwing Inc., Homewood (1979)

7. Keegan, W.: Global Marketing Management, 4th edn. Prentice-Hall International Editions, Englewood Cliffs (1989)
8. Dahringer, L., Mühlbacher, H.: International Marketing, a global perspective. Addison-Wesley Publishing company, Reading (1991)
9. Hennessey, J.-P., Hennessey, D.: Global Marketing Strategies, 5th edn. Houghton Mifflin Company, USA (2001)
10. McLuhan, M.: Understanding media: The extensions of man. McGraw-Hill, New York (1964)
11. de Mooij, M.: Convergence and divergence in consumer behaviour: implications for global advertising. International Journal of advertising 22, 183–202 (2003)
12. de Mooij, M., Hofstede, G.: Convergence and divergence in consumer behavior: implications for international retailing. Journal of Retailing 78, 61–69 (2002)
13. Davis, F.D.: Perceived usefulness, perceived ease of use and user acceptance of information technology. MIS Quarterl 13(3), 319–340 (1989)
14. Yoh, E., Damhorst, M.L., Sapp, S., Laczniac, R.: Consumer adoption of the Internet: The case of apparel shopping. Psychology & Marketing 20(12), 1095–1118 (2003)
15. Constantinides, E., Geurts, P.: The impact of web Experience on virtual buying behavior: An empirical study. Journal of Customer Behavior 5, 307–336 (2006)
16. Hofstede, G.: Cultures and Organizations: Software of the Mind. McGraw-Hill, New York (1997)
17. Hofstede, G.: Culture's consequences: International differences in work-related values. Sage, Newbury Park (1980)
18. Hofstede, G.: Culture's consequences, comparing values, behaviours, institutions, and organizations across nations. Sage Publications, Thousand Oaks (2001)

Evolutionary Prediction of Online Keywords Bidding

Liwen Hou, Liping Wang, and Jinggang Yang

535, Fahuazhen Rd., Department of Management Information System
Shanghai Jiaotong University, Shanghai, China

Abstract. Online keywords bidding as a new business model for search engine market facilitates the prosperity of internet economy as well as attracts myriad small business to directly target their customers. In order to effectively manage each advertising campaign the manager needs to figure out smart strategy to bid. This paper successively developed two models (static and dynamic) with four crucial variants at keyword bidding (bid, rank, click through and impression) based on the bayesian network to assist the prediction of bidding. Herein dynamic model, evolved from the static model, takes the influence of click through on the last period into account and extends the strategy space. Empirical study by aid of data from the largest online travel agency in China is carried out to test both models and the results indicate that they are effective while the dynamic model is more attractive in terms of prediction accuracy. Finally further research directions along this paper is shown.

Keywords: Dynamic Bayesian Network, Keyword Bidding, Click Through.

1 Introduction

As the giant in search engine market Google earns much money from keywords auction, well appeared as couple of sponsored links listed in the right side of the searched pages. Many others quickly get involved afterwards and ignite this industry. Since being able to provide a feasible approach to advertising through internet while against the expensive traditional media and assumed to directly target customers in favor of those smaller and unknown companies sponsored link Ad does give an edge to business and has boomed since then, just as seen in Fig.1.

Myriad business resorts to sponsored link to get known, even larger enterprises who perceived this synthetic power and advantages to combine online marketing with physical campaign also consider to capitalizing on part of budget to achieve it. Advertisers, however, firstly exited about this new channel and now got confused in the sense that they have no chance to know in detail the bidding process as well as the consumption of their Ad budget, which may stymie the evaluation to their advertising alternatives. In addition there typically exists a many-to-many relationship between the keyword set and advertiser set. That is, each of the majority of keywords will be bided by several advertisers who in turn manage a keyword portfolio. As a result any intention to occupy the highest position in the sponsored links is doomed to exhaust out the Ad budget much quickly and armed with strong incentive to explore lower position. Such extreme asymmetric game definitely aggravates the existence and undulation of the instant equilibria, which pushes advertisers trapped in continuous

G. Psaila and R. Wagner (Eds.): EC-Web 2008, LNCS 5183, pp. 124–133, 2008.

and irrational bidding. For example a generic keyword "flower" was bided from ¥ 0.21 to ¥ 13.05 for the first position on Google during Dec.2006 in China because 96 companies got involved and triggered off intense competition at that time. Many advertisers afterwards complained that they harvested less than they paid out for this campaign due to lacking the optimal budget management. Consequently the most urgent need for advertisers is to figure out some strategy to bid smartly and avoid suffering as much as possible otherwise the "stupid money" will still stick to them in case of mismanaged campaigns.

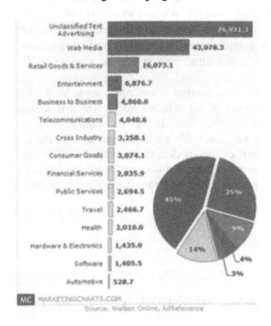

Fig. 1. Sponsored link by industries (11.2007)

At present scholars have exploited this field by different methodologies but still in its infancy so related literature is mostly very recent. Research by Animesh et al (2005) considered different profiles of bidding strategy, and examined the underlying reasons for the customers' quality signaling under a search-experience-credence framework. They also examined the relationship between advertisers' quality and their bidding strategies in online setting. Benjamin et al. (2005) aimed at analyzing the strategic behavior of keyword auctions. They found strategic behavior impaired the benefit of both search engine and market. Feng et al. (2005) inspected four kinds of mechanisms with respect to the correlation of advertiser's willingness to pay and term relevance in the steady state setting and two alternative approaches under dynamic situation. Brendan et al (2004) simulated PPC (pay per click) keyword auction based on intelligent agent technology. A series of rules are provided to the agent imitating bidders and the real bidding process is illustrated in the Overture platform (now part of Google) given such exogenous variables as profit function. Agent technology was also used by Sunju et al. (2004), who elaborately induced agents to intelligently bid using a Markov chain model. They also identified the optimal bidding strategies in an experiment by combining those parameters addressing the auction situation in a given testbed. Young-Hoon (2005) proposed an integrated framework to simulate bidding behavior for keyword auctions on a website. They concluded that out of the four key components willingness to bid is the only significant factor.

This paper will address the prediction of online keywords bidding using bayesian network (BN) technology in an evolutionary sequence from the static to dynamic model according to influencing relationship of bidding profile. The contributions of the paper can be unfolded as the following three aspects: method, evolution view and empirical test. BN is rarely applied to online keyword auction in previous research but does function as an effective vehicle to reflect the causal linkage among such four core variants of keyword bidding strategy as *bid, rank, click_through* (abbr. *CT*) and *impression*. On the other hand Google put on new policy in 2005 to facilitate the promotion of online advertising for small business, which led to the extension of bidding prediction from only one period to successive two periods conditional on new parameters to be learned simultaneously supplemented. This evolution of prediction approximates the reality more by taking dynamics into account so that advertisers can lay down the budget in a relative long term. In order to verify the prediction models this paper collected data from the largest online travel agency in China to test them. The results show that our models are valid and attractive and may be adopted by advertisers with little modification.

The rest of the paper is organized as follows: Section 2 develops a static prediction model based on BN and then addresses types of the selected nodes which represent those parameters (or variants) mentioned above. Considering the request to cover sequential campaigns Section 3 develops a dynamic BN prediction model with the aid of new learning functions. Section 4 verifies our models using empirical data collected from the largest online travel agency in China and the results are proved to be acceptable. Finally, Section 5 summarizes the paper and shows the further research direction.

2   Static Prediction

2.1   Methodology

BN as one of the advanced modeling methods representing the joint probability distributions of variants (i.e. nodes in the network) is consisted of structure model and parameter model. The former, based on a diagram, illustrates the interactive relationship (most likely being causal link) among those variants while the latter addresses the conditional probabilities of variants given their parent nodes. The technology is rather powerful due to couple of potentials, such as multifactor input and prior knowledge or experiences involved. The details can be found in reference [8]. Towards this end people often employ bayes technology in artificial intelligence to assist uncertain reasoning in different domains like medical diagnosis and heuristic search. For online bidding it is also applicable because not only the causal links between these variants need to be simultaneously described but also the bidding history is an important prior for the subsequent bidding strategy. In this paper BN is associated with system modeling for decision, a generic application pro tango.

In general the construction of a BN involves three steps: ①identifying the dominant variants; ②ascertaining interdependent and independent relationships among those variants; ③learning and determining probability parameters for the existing relationship. Herein we will only concentrate the last step because it is the toughest analysis and needs more consideration.

2.2 Model

In online keyword auction *bid*, *rank*, *CT* and *impression* are four the most essential variants with respect to bidding strategy. Herein *impression* refers to appearing times resulted from searching. *Bid* and *rank* have obvious causal link while the *rank* is the cause of *CT* as well if the assumption that higher *rank* brings about more clickstream held. In addition *CT* will change with *impression* in the sense that the number of search of one keyword accrued in some period given *CT* = *impression* * *CT_Rate*. Assuming the higher rank the keyword the more possible to click it we can construct a directed acyclic graph with these four variants shown in Fig.2, which implies: *rank* = *f* (*rank* * *CT*) and *CT=g* (*rank*, *impression*). Both functions can further scratch up an implicit probability listed in the following:

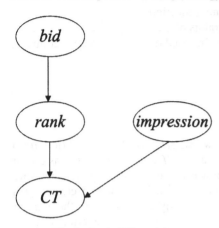

Fig. 2 . Static BN model

$P(bid, rank, impression, CT) =$
$P(bid)*P(rank \mid bid)*P(impression)*P(CT \mid rank, impression)$

On the other hand total probability formula tells that:
$P(bid, rank, impression, CT) =$
$P(bid)*P(rank \mid bid)*P(impression)*P(CT \mid bid, rank, impression)$

Both formulae above show the following relationship
$P(CT \mid bid, rank, impression) = P(CT \mid rank, impression)$

which means *CT* is independent to *bid* given *rank*, congruent with the reality.

2.3 Parameter Learning

Just as asserted previously that parameter learning is rather challengeable for bayesian prediction and the prerequisite to do that is firstly make clear the type (i.e. discrete, continuous and mixed) of nodes inasmuch as different types are associated with different learning algorithms. We choose the mixed type for all nodes because *bid* and *rank* can be viewed as the discrete variants (also called Tabular nodes) and *impression* and *CT* as the continuous (also called Gaussian nodes). This judgment is a little arbitrary though, it is acceptable if observing change frequency of each variant in fixed period.

The Essential of parameter learning based on history data is to adjust the conditional probabilities associated with the four variants. Therefor the complete maximum likelihood estimate (MLE) needs to be employed to address the learning in the presence of existed data set. Herein a powerful Toolbox BayesNet coded by Kevin

Murphy is used to carry on the prediction process, where five functions shown in the following are involved:

mk_bnet: establishment of structure of BN

bnet.CPD{i}: assign the type to the node i in BN

learn_params: parameter learning function

enter_evidence: variant constraints of reasoning

jtree_inf_engine: inference engine used during parameter learning

3 Dynamic Prediction

3.1 Quality Score

This paper technically aims at Google auction whose ranking algorithm differs from other search engines like Yahoo due to incorporating *CT* rate (abbr. CTR) instead of completely bidding oriented. This restriction, however, ruled out those websites with low CTR appearing on the top position of right side. In 2005 Google changed his ranking algorithm so that part of the websites previously excluded from the anterior searched pages is reignited to strive for favorable positions. This is the effect of quality score defined by Google as the base to evaluate keyword quality and decide lowest bidding. Besides CTR quality score is related with Ad relevance, history records of keyword effect and other minor factors, among which history effect plays the fundamental role in our model because this impact results in a dynamic situation. Towards this point the static model evolves into the dynamic one, inheriting previous performance and allows trial and error.

3.2 Model

The introduction of quality score oscillates the assumption about static model that *rank* is only affected by *bid*. Instead history *CT* of a keyword also heralds such influence. This modification not only brings out a dynamic model of keyword bidding but also allows advertisers to allocate Ad budget more reasonable in long period. This dynamic bayesian model, shown as Fig.3, is developed by incorporating dynamic mechanism into the static model. Dynamic BN as a schematic representation to the complex stochastic process arguments BN by modeling the process of time sequence and the change with time of the stochastic variants set. In this sense dynamic bayesian model paves a path to explore the prediction of keyword bidding with respect to evolutionary view, a more realistic approach for advertisers.

3.3 Parameter Learning

Similar to that of static model dynamic model needs to learn those conditional probabilities as well. Nevertheless a difference between both is the type of nodes. Herein node *CT* and *impression* in the previous phase is considered as the parent node of *rank* and should be labeled as discrete because mixed learning algorithm is still under mature and hence inapplicable to our model. Therefore all nodes in this dynamic model are attributed to discrete (i.e. tabular) type.

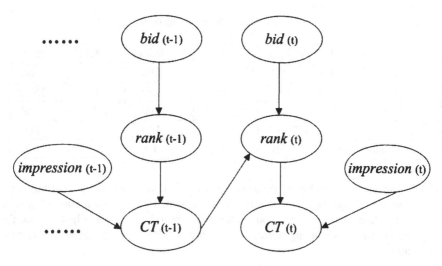

Fig. 3. Dynamic BN model

An accompanying issue to this node treatment is to discretize the continuous variants *CT* and *impression*. Therefor one should firstly select the number of values of continuous variants (i.e., those values can mimic the shape of the variant) to be discretized and then segment the continuous interval corresponding to some period to guarantee each subinterval encompassing one value of one of the continuous variants. Thus the variant can be represented by those discretized values if the number of subintervals enough. Several algorithms such as equal width interval and weighted information-loss discretization can serve for this purpose. This paper chooses the former algorithm, the simplest one because of no category information about type needed. In addition we assume all conditional probabilities of each variant obey uniform distribution so that they can be adjusted adequately by training.

In order to realize the dynamic bayesian model three more functions in the Bayes-Net Toolbox are involved and shown as the following

mk_dbn: establishment of dynamic structure of BN
dbn.CPD{i}: assign the type to the node i in dynamic BN
jtree_2TBN_inf_engin: dynamic inference engine of two-period interaction

4   Case Analysis

4.1   Data Description

In order to test the validity and effect of the models developed above we luckily acquired the experimental data from the largest online travel agency in China. The dataset, spanning from Dec.21 to Dec.28 2007 on Google platform, includes 299 time slices, each of which tracks such information as bidding time, average bidding, click through, rank and so on. The statistics of the dataset is summarized in the Tab.1.

Table 1. Statistics of bidding dataset

Variants	Average	S.t.	Median	Min	Max
bid	1.01	0.55	1.01	0.01	2.00
rank	8.91	3.47	8.84	1.56	15.77
impression	3005.47	328.00	3038	2074	3956
click through	222.14	90.61	196	46	478
conversion rate	22.26	11.12	19	5	63

Analyzing the relationship of those variants in the dataset we found that the number of search of single keyword within a time slice will change with slice sequence. That is, the slice close to right end incurs more searching times. Fig.4 shows the test result of *impression* using normplot function in Matlab to analogize the normal distribution.

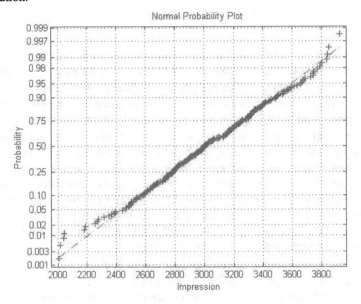

Fig. 4. Normal distribution test of *impression*

Suppose the more the data points juxtapose a line the more likely the distribution of the variant is paralleled to normal distribution. So the searching times of any single keyword, whatever *impression* or other variants, can be assumed to obey normal distribution. Actually we experimented *rank* and obtain the same result.

4.2 Prediction Results

In the presence of structure model of BN shown in Fig.2 and Fig.3 we split the sample dataset into two parts. The first part including 200 time slices serves as the training set which feeds data to the BN models and all parameters (i.e. a series of conditional probabilities) will be computed using the functions in the BayesNet mentioned above. The process is iterative so that the final results can converge to some value. The second part consisted of the rest of 99 slices acts as the tester to make sure the predicted results acceptable.

Fig.5 shows the true value (from the 99 time slices given P(*rank* | *bid*)), predicted value (from our model) and their gap with respect to *rank* from top to down when the static model works.

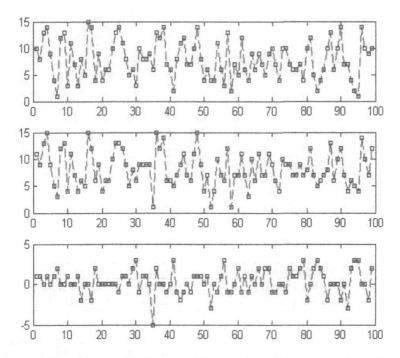

Fig. 5. True value, predicted value and their gap for *rank* based on static model

Similarly we can obtain the schemas of other three variants using this static bayesian model. While the dynamic model roughly remains the same process dynamic inference engine is embedded and works for the learning. Fig.6 shows the predicted result with respect to *rank* as well.

Intuitively dynamic bayesian model seems to be more accurate than its counterpart because more zero points appears in the gap diagram if glimpsing the graph. Actually it is true when the general statistics associated with both predictions above are compared. See Tab.2.

Obviously dynamic model indicates smaller deviation with respect to predicted mean and standard error in terms of variant *rank*. This means dynamic model not only evolves to be more desirable and effective for keyword bidding management but also proves that *CT* in the previous round does influence the *rank* in this round, as shown

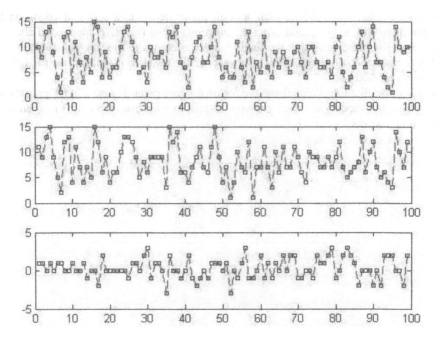

Fig. 6. True value, predicted value and their gap for *rank* based on dynamic model

Table 2. Comparison of rank prediction of two models

Deviation statistics	Static	Dynamic
Predicted mean	0.30303	0.213131
Predicted standard error	1.48086	0.91434

in Fig.3. Notice that during the prediction the keyword remains thoroughly coincident while the searching times in the meantime obey normal distribution. So the prediction error can be further dwindled once the information about searching times of a specific keyword acquired.

5 Conclusion

The importance and attractiveness of online marketing have been realized and fulfilled for couple of years. As the frontier of the practice sponsored link provides effective and efficient channel for business to approach customers. The fact that billions of dollars swarm into this market facilitates the prevalence of keywords auction as well as stimulates the prosperity of Internet economy. However myriad advertisers relapsed into an adverse state because of ignorance to bidding strategies and competitors behavior. This paper developed two models based on BN to help advertisers predict such crucial variants as *bid*, *rank*, *CT* and *impression* in online keyword auction. The results indicate that *CT* in the previous round does influence *rank* in the current round and hence dynamic bayesian model exceeds the static model in terms of prediction precision. The contributions of this paper lie in the introduction of evolutionary prediction from static to dynamic model by incorporating both knowledge on history bidding and impact of previous click and meanwhile the empirical study by aid of data from the largest online travel agency in China.

Definitely this research can be enhanced from two aspects. One is the inclusion of more variants to dynamic model. Since the history values of *bid, rank* and *impression* instead of only *CT* all may influence the *rank* prediction of current round the model can evolve to be more complete by assigning more parent nodes to rank. Another improvement results from the attribution of keywords. Travel and ticket are the examples of perishable goods which have obvious expiration so their inventory level will influence their bidding while it may escape such issue in the case of imperishable industrial goods. So managers may tailor different bidding strategies for different goods. Summarily we expect further research to underpin those aspects above.

References

1. Animesh, A., Ramachandran, V., Viswanathan, S.: Online Advertisers Bidding Strategies for Search, Experience, and Credence Goods: An Empirical Investigation. Working paper (2005)
2. Animesh, A., Ramachandran, V., Viswanathan, S.: Quality Uncertainty and Adverse Selection In Sponsored Search Markets, working paper (2005)
3. Edelman, B., Ostrovsky, M.: Strategic Bidder Behavior in Sponsored Search Auctions. Int. J. Internet Marketing and Advertising 2, 92–108 (2005)
4. Kitts, B., Leblanc, B.: Optimal Bidding on Keyword Auctions. Electronic Markets, special issue: innovative auction markets 14, 186–201 (2004)
5. Feng, J., Bhargava, H.K., Pennock, D.M.: Implementing Sponsored Search in Web Search Engines: Computational Evaluation of Alternative Mechanisms. working paper (2005)
6. Park, S., Durfee, E.H., Birmingham, W.P.: Use of Markov Chains to Design an Agent Bidding Strategy for Continuous Double Auctions. Journal of Artificial Intelligence Research 22, 175–214 (2004)
7. Park, Y.-H., Bradlow, E.T.: An Integrated Model for Bidding Behavior in Internet Auctions: Whether, Who, When, and How Much. Journal of Marketing Research XLII, 470–482 (2005)
8. Jensen, F.V.: Bayesian Networks and Decision Graphs. Springer, Heidelberg (2001)
9. Murphy, K.: Bayes Net Toolbox for Matlab (Accessed November 15, 2007) (2007), http://www.cs.ubc.ca/~murphyk/Software/BNT/bnt.html
10. Kitts, B., Leblanc, B.: Optimal Bidding on Keyword Auctions. Electronic Markets 14, 186–201 (2004)

Web Behaviormetric User Profiling Concept

Peter Géczy, Noriaki Izumi, Shotaro Akaho, and Kôiti Hasida

National Institute of Advanced Industrial Science and Technology (AIST)
Tokyo and Tsukuba, Japan

Abstract. We present a concept for building behaviorally centered user profiles. The concept utilizes behavioral analytics of user interactions in web environments. User interactions are temporally segmented into elemental browsing units. The browsing segments permit identification of the essential navigational points as well as higher order abstractions. The profiles incorporate relevant metrics from three major domains: temporal, navigational, and abstractions. Temporal metrics focus on aspects of durations and delays between portions of human interactions. The navigational metrics target the initial, terminal, and single user actions. The abstraction metrics encompass elemental patterns of human browsing behavior and their interconnections. The profiling concept utilizes relatively simple analytic and statistical apparatus. It facilitates computational efficiency and scalability to large user domains.

1 Introduction

The demand for personalized approaches to interaction in web environments is rapidly increasing. Conventional one-fit-all designs in web commerce have been facing significant usability issues. Modern web commerce requires flexible non-disruptive personalization techniques [1],[2]. Personalized systems benefit both users and organizations.

The key to personalization is efficient profiling of users. The information in profiles is used for customizing the web interactions, sales, or marketing to individual users [3]. Creation of profiles necessitates collecting and analyzing the individual's use characteristics [4],[5]. Information stored in profiles may vary depending on the application domain. Conventional approaches to personalization have been history and content-based [6]. Researchers and practitioners have been largely utilizing the historical records of user navigation (or searches) interlinked with the processed content of accessed resources. Modern interactive web presence requires personalization approaches embracing more detailed elucidation of human interactions with the system.

We introduce a novel profiling concept addressing the needs of modern web commerce for user behavioral analysis. The presented concept is based on the detailed analysis of human-web interactions. The analysis enables derivation of effective behavioral metrics suitable for individual and group profiles. The metrics span throughout temporal, navigational, and pattern abstraction domains.

G. Psaila and R. Wagner (Eds.): EC-Web 2008, LNCS 5183, pp. 134–143, 2008.

2   Conceptual Approach

User profiling is generally a multi-stage process. Three essential stages of profiling process are: data collection, (pre-)processing, and profiling. The flow is basically sequential (as illustrated in Figure 1), however, the process cyclically repeats whenever new data is collected. Computational intensity of each stage varies depending on the scale of user base and activity. Small scale data may be processed on-the-fly, whereas voluminous data may require significantly longer processing. Conventionally, the user profiles are updated periodically; e.g. overnight, so the following day user profiles already account for yesterday's activity.

Fig. 1. Depiction of profiling process consisting of three general steps

Data Acquisition. Information about user interactions in web environment can be collected variously. Depending on the collection point, one can distinguish two main categories: server-side and client-side data acquisition. The server-side data acquisition refers to the web logs produced by a web server. Web servers have capability to record information about each arriving request and served object. The information is stored in a predefined format. The client-side data acquisition is performed by software agents executed locally on user's machine. The agents can be standalone programs (running in the background), client-side scripts, or directly web browsers. They collect required information and forward it to the data collection server.

Server-side and user-side data acquisition methods must allow *user identification*. The user identification can be direct or indirect. The direct identification uniquely identifies a particular user and is more accurate (e.g. required unique login associated with cookies and session identification information). The indirect identification is less accurate and essentially relates to the origin of the activity (e.g. computer with dynamically assigned IP address that can be used by several users). The set of users $U = \{u_1, \ldots, u_n\}$, in practice, may be identified by both direct and indirect means. Identification of users is inevitably linked with privacy issues and concerns.

Processing. Acquired data about user interactions often requires processing before it can be used for creating profiles. The recorded data by web servers, for instance, contains logs of both human and machine generated traffic. The machine traffic should be suitably identified and eliminated. The data, even after elimination of machine generated traffic, may still contain records on served objects that are irrelevant for user profiling (e.g. stylesheet files of web pages). Only information relevant to profiling should be kept. The extracted data may require additional specialized processing. The specialized processing aims at delivering 'raw' information utilized for profiling.

Profiling. The (pre-)processed data containing information about user web interactions is utilized for extracting a set of relevant features $\Phi = \{\phi_1, \ldots, \phi_m\}$. The relevant features can be determined *a priori*, or dynamically by applying data mining and/or pattern detection techniques. Having a set of users $U = \{u_1, \ldots, u_n\}$ and features $\Phi = \{\phi_1, \ldots, \phi_m\}$, the profiling is a mapping $P : U \to \Phi$ where each user u_i is associated with a set/vector of features.

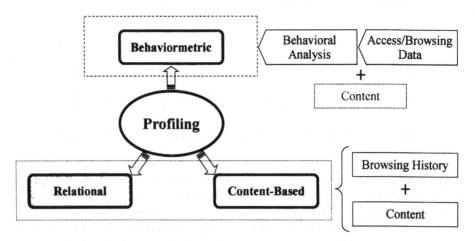

Fig. 2. Conceptual categorization of profiling techniques

Conventional profiling approaches can be categorized into two main groups: content-based and relational (Figure 2) [7]. Both approaches utilize browsing history and content of accessed pages for creating profiles. The content-based methods associate detected users with content elements of visited pages during their browsing. The elements of the content can range from simple text to multimedia objects. These approaches have been largely explored by search engines for marketing purposes. User profiles, for instance, contain the searched keywords and words extracted from the accessed pages together with related metrics. The keyword profiles are used for matching textual advertisements displayed to the users during their interactions with search engine (or other pages serving the advertisements).

The relational methods go a step further by employing the 'similarity' concepts. Similarity measures between various objects can be defined, e.g. between simple elements such as words and pages, more complex ontologies or semantic representations, and even the whole profiles [8]. They quantitatively express a level of inter-relatedness. The relational profiling incorporates links between inter-related objects. These profiling techniques have been utilized for collaborative filtering and recommendations in electronic commerce systems. For example, when a user expresses interest in the certain item, other similar/related items may be offered, too (or items purchased by 'likeminded' users having similar profiles or purchasing characteristics).

Behaviormetric Profiling. The profiling concept presented in this study is based on the detailed elucidation of user browsing behavior and appropriate metrics (Figure 2). The behaviormetric profiling addresses significantly more fundamental issues than just a simple account of browsing history and content of visited pages. Naturally, the behaviormetric profiling can be extended for the content based analytics and relational links, however, preferably at later stages. The core metrics are derived from the analysis of human browsing behavior.

2.1 Browsing Analytics

Human dynamics in electronic environments have been observed to have temporally specific features [9]. Users display periods of activity followed by longer periods of inactivity. Thus the human-web interactions can be segmented with respect to the temporal indicators. The temporal segmentation framework of human browsing behavior has been introduced in [10]. We concisely recall constructs relevant to this study.

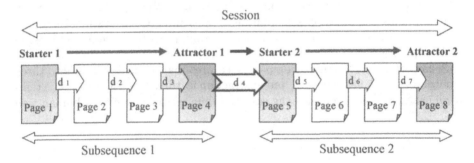

Fig. 3. Segmentation of web interactions and identification of navigation points

The sequences of page transitions are segmented into *sessions* and *subsequences*. The segmentation is illustratively depicted in Figure 3. The sessions outline tasks of various complexities undertaken by the users. They are further divided into the subtasks represented by the subsequences. Consider the sequence of the form: $\{(p_i, d_i)\}_i$ where p_i denotes the visited page URL_i and d_i denotes a delay between the consecutive views $p_i \rightarrow p_{i+1}$. **Session** is a sequence $B = \{(p_i, d_i)\}_i$ where each $d_i \leq T_B$. **Subsequence** $S = \{(p_i, d_i)\}_i$ consists of pairs (p_i, d_i) with delays satisfying the conditions: $d_i \leq T_S$ and $\{(p_i, dp_i)\}_i \subset B$.

There are several important navigation points: the points where users initiate their browsing actions—*starters*, the resources they target—*attractors*, and single action behaviors—*singletons*. **Starters** are the first points of subsequences or sessions with length greater that 1. **Attractors** are the last points of sub-sequences or sessions with length greater that 1. **Singletons** are the points of subsequences or sessions with length equal to 1.

Subsequences are the elemental browsing segments in which users navigate from the initial point to the desired target. Users can take multitudes of navigational paths from the initial navigation point to the target. Since the points between the starter and the attractor are essentially transitional, keeping the history of all navigational pathways may be a waste of resources. It is desirable to create an abstract representation of browsing segments. Starter-attractor pairs, denoted as **SA elements**, serve as suitable abstractions of subsequences. The transitions between subsequences are represented by attractor-starter pairs (attractor of one subsequence and starter of the consecutive subsequence). They are referred to as **connectors**, because they connect the subsequences. Formation of more complex browsing patterns can be observed from sequences of SA elements and connectors.

3   Behaviormetric Profiling Framework

Introduced browsing analytics present valuable data about interactions of users with the web environment. They provide the essential information base. The behavioral metrics are derived from the browsing analytics data by employing relatively simple statistical and analytical apparatus. Computational simplicity facilitates scalability of behaviormetric profiling to the systems with large number of users.

Behaviormetric profiling encompasses features that are categorized into three groups: temporal, navigational, and abstraction features (see Figure 4). The temporal characteristics underline the essential time related specifics of human-web

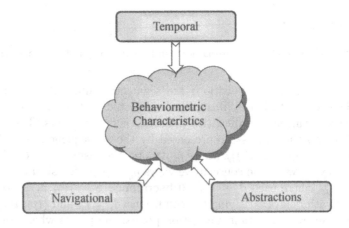

Fig. 4. Three main domains contributing to the behaviormetric profiling features

interactions. The characteristics are derived from detected durations of sessions and subsequence, as well as transitional delays between page views and subsequences. The navigational profiling characteristics expose the aspects of user interactions related to the important navigational elements such as starters, attractors, and singletons. The abstraction characteristics highlight repetitive elemental and complex browsing pattern formations based on observed SA elements and connectors data.

The presented behaviormetric concept is supported by illustrative typical real-world characteristics. The characteristics were obtained from the behavioral analytics at a large-scale organizational web portal of The National Institute of Advanced Industrial Science and Technology. The portal is significantly complex in terms of the infrastructure, resources, and user base.

3.1   Real-World Intranet Portal

The portal infrastructure consists of six web servers connected to the high-speed backbone in a load balanced configuration. The accessibility of intranet services and resources to the users is provided via connectivity ranging from optical to wireless. Supported platforms extend to mobile devices.

A wide spectrum of resources on the portal include documents in various formats, downloadable software, multimedia, etc. There are also hundreds of services supporting organizational business processes, cooperation with industry, academia, and other institutes, resource localization; but also networking, blogging, etc. Visible web space exceeds 1 GB, and deep web space is considerably larger.

Significant traffic generates rich pool of web log data. The data preparation, processing, filtering, and segmentation to sessions and subsequences are described in [10] and not addressed here. Concise data description is in Table 1.

Table 1. Recorded one year web portal interaction data

Log Records	315 005 952
Resources	3 015 848
Sessions	3 454 243
Unique Sessions	2 704 067
Subsequences	7 335 577
Unique Subsequences	3 547 170
Valid Subsequences	3 156 310
IP Addresses	22 077

3.2   Temporal Characteristics

The underlying temporal characteristics are represented by delays and durations. The typical pattern of temporal characteristics is shown in Figure 5. It depicts part of the histogram of average subsequence durations in sessions. The histogram has roughly Poissonian-like distribution.

Three distinct segments of the distribution are noticeable. The initial part indicates short subsequences where the transitions between page views are rapid. The middle part contains the subsequences with intermediate duration. Majority of sessions contain these subsequences. The third segment denotes longer lasting subsequences. Other delay and duration characteristics generally follow the mentioned features. Thus the essential temporal aspects of user behaviormetric profiles are suitably represented by minimum, maximum, and average diagnostics:

Durations: representing time lengths–denoted by l;
  - *minimum* durations for sessions, $l_{B_{min}}$, and subsequences $l_{S_{min}}$,
  - *maximum* durations for sessions, $l_{B_{max}}$, and subsequences $l_{S_{max}}$,
  - *average* durations for sessions, $l_{B_{avrg}}$, and subsequences $l_{S_{avrg}}$.

Delays: representing delay lengths–denoted by d;
  - *minimum* delays between sessions, $d_{B_{min}}$, and subsequences $d_{S_{min}}$,
  - *maximum* delays between sessions, $d_{B_{max}}$, and subsequences $d_{S_{max}}$,
  - *average* delays between sessions, $d_{B_{avrg}}$, and subsequences $d_{S_{avrg}}$.

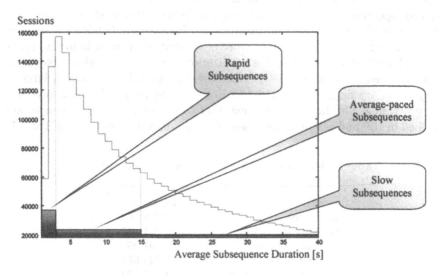

Fig. 5. Typical temporal characteristics of average subsequence durations in sessions

3.3   Navigational Characteristics

The navigational characteristics target the essential observations related to three important categories of navigation points: starters, attractors, and singletons. The typical access pattern for these navigation points is displayed in Figure 6. It shows histogram and quantile aspects of starters access. Similar characteristics hold also for attractors and singletons.

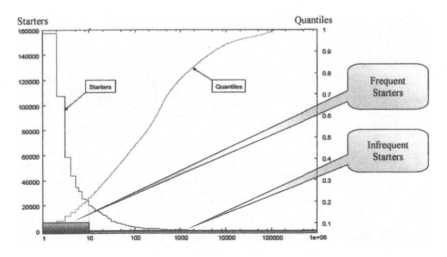

Fig. 6. Typical histogram and quantile characteristics of starting navigation points

An evident long tail distribution is exposed in the histogram chart (note that x-axis is in logarithmic scale to provide better view of the head of distribution). The narrow head contains relatively small number of frequently accessed points. The extensive tail contains large number of infrequently accessed points. It is pertinent that user profiles encompass the manifested observations. This is done by means of frequencies, numbers of unique elements, and selected number of the most frequent elements:

Top-x Sets: containing the most frequently accessed x number of starters $(T_S^{(x)})$, attractors $(T_A^{(x)})$, and singletons $(T_Z^{(x)})$.

Number of Unique Elements: quantifying the number of unique starters, attractors, and singletons accessed.

Relative frequencies: representing relative access statistics–denoted by r;
  - *Starters:* relative frequencies r_{S_i} for all starters S_i in $T_S^{(x)}$,
  - *Attractors:* relative frequencies r_{A_i} for all attractors A_i in $T_A^{(x)}$,
  - *Singletons:* relative frequencies r_{S_i} for all singletons Z_i in $T_Z^{(x)}$.

3.4   Abstraction Characteristics

The abstractions characteristics emphasize higher order elemental and complex browsing interaction patterns expressed by SA elements and connectors. The representative pattern for connector utilization is shown in Figure 7. It displays both histogram and quantile charts. SA elements exhibit similar features.

The histogram clearly exposes long tailed characteristics analogous to those of navigation points. Relatively small numbers of SA elements and connectors are frequently used; with significantly larger numbers of them used occasionally.

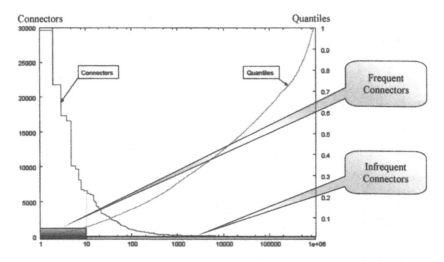

Fig. 7. Typical histogram and quantile characteristics of subsequence connectors

Frequently used elemental patterns (represented by SA elements) interconnected by frequent transitions uncover formation of more complex interaction patterns. Profiling abstractions is more efficient that tackling large number of different complex patterns. Analogously to navigational characteristics, the abstraction profiles incorporate relative frequencies, unique number of abstractions, and the most frequent abstraction sets:

Top-x Sets: containing the most frequently utilized x number of SA elements $(T_{SA}^{(x)})$ and connectors $(T_C^{(x)})$.

Number of Unique Abstractions: quantifying the number of unique SA elements and connectors utilized.

Relative frequencies: representing relative utilization statistics, r;
  - *SA elements:* relative frequencies r_{SA_i} for all SA elements SA_i in $T_{SA}^{(x)}$,
  - *Connectors:* relative frequencies r_{C_i} for all attractors C_i in $T_C^{(x)}$.

4   Conclusions

We presented a novel approach to user profiling. The approach is based on the fundamental analysis of human interactions in web environments. Analysis of human web behavior utilizes segmentation of browsing interactions. The behavioral segments enable identification of the essential navigational points as well as higher order abstractions. The higher order abstractions outline the elemental browsing and transitional patterns. The introduced profiling concept envelops three pertinent domains: temporal, navigational, and higher order abstraction characteristics. The temporal characteristics contain durations and delays specific to browsing segments. The navigational characteristics encompass the intail and terminal navigation points, as well as single user actions. The abstraction characteristics expose the elemental browsing patterns and transitions between them. The profiling concept employs simple and computationally inexpensive statistical and analytic apparatus. This facilitates the scalability of the concept to the large user domains.

References

1. Baraglia, R., Silvestri, F.: Dynamic personalization of web sites without user intervention. Communications of the ACM 50, 63–67 (2007)
2. Mobasher, B.: Data mining for web personalization. In: Brusilovski, P., Kobsa, A., Nejdl, W. (eds.) The Adaptive Web, pp. 90–135. Springer, Heidelberg (2007)
3. Moe, W.W.: Buying, searching, or browsing: Differentiating between online shoppers using in-store navigational clickstream. Journal of Consumer Psychology 13, 29–39 (2003)
4. Nasraoui, O., Soliman, M., Saka, E., Badia, A., Germain, R.: A web usage mining framework for mining evolving user profiles in dynamic web sites. IEEE Transactions on Knowledge and Data Engineering 20(2), 202–215 (2008)
5. Gauch, S., Speretta, M., Chandramouli, A., Micarelli, A.: User profiles for personalized information access. In: Brusilovski, P., Kobsa, A., Nejdl, W. (eds.) The Adaptive Web, pp. 54–89. Springer, Heidelberg (2007)
6. Gasparetti, F., Micarelli, A.: Exploiting web browsing histories to identify user needs. In: Proceedings of the 12th International Conference on Intelligent User Interfaces, New York, NY, USA, pp. 325–328 (2007)
7. Adomavicius, G., Sankaranarayanan, R., Sen, S., Tuzhilin, A.: Incorporating contextual information in recommender systems using a multidimensional approach. ACM Transactions on Information Systems 23, 103–145 (2005)
8. Anand, S.S., Kearney, P., Shapcott, M.: Generating semantically enriched user profiles for web personalization. ACM Transactions on Internet Technology 7(4), 22 (2007)
9. Barabasi, A.-L.: The origin of bursts and heavy tails in human dynamics. Nature 435, 207–211 (2005)
10. Géczy, P., Akaho, S., Izumi, N., Hasida, K.: Knowledge worker intranet behaviour and usability. Int. J. Business Intelligence and Data Mining 2, 447–470 (2007)

Author Index

Lecture Notes in Computer Science

Sublibrary 3: Information Systems and Application, incl. Internet/Web and HCI

For information about Vols. 1– 4777
please contact your bookseller or Springer